Springer Climate

Series editor

John Dodson, Menai, Australia

Springer Climate is an interdisciplinary book series dedicated on all climate research. This includes climatology, climate change impacts, climate change management, climate change policy, regional climate, climate monitoring and modeling, palaeoclimatology etc. The series hosts high quality research monographs and edited volumes on Climate, and is crucial reading material for Researchers and students in the field, but also policy makers, and industries dealing with climatic issues. Springer Climate books are all peer-reviewed by specialists (see Editorial Advisory board). If you wish to submit a book project to this series, please contact your Publisher (elodie.tronche@ springer.com).

More information about this series at http://www.springer.com/series/11741

Katharine Jacobs • Susanne Moser • James Buizer
Editors

The US National Climate Assessment

Innovations in Science and Engagement

Previously published in *Climatic Change* Volume 135,
Issue 1, 2016

Springer

Editors
Katharine Jacobs
Institute of the Environment
University of Arizona
Tucson, Arizona, USA

Susanne Moser
Susanne Moser Research and Consulting
Santa Cruz, California, USA

James Buizer
Institute of the Environment
University of Arizona
Tucson, Arizona, USA

ISSN 2352-0698 ISSN 2352-0701 (electronic)
Springer Climate
ISBN 978-3-319-41801-8 ISBN 978-3-319-41802-5 (eBook)
DOI 10.1007/978-3-319-41802-5

Library of Congress Control Number: 2016949492

Printed on acid-free paper

This Springer imprint is published by Springer Nature
The registered company is Springer International Publishing AG Switzerland

Contents

Climatic Change (2016) 135:1–7
DOI 10.1007/s10584-016-1621-5

The third US national climate assessment: innovations in science and engagement

Katharine L. Jacobs[1] **· James L. Buizer**[2] **·**
Susanne C. Moser[3,4]

Received: 15 January 2016 / Accepted: 21 January 2016 / Published online: 10 February 2016

1 Introduction

Climate change poses numerous challenges for ecosystems, communities, businesses, and government agencies, and these challenges are becoming more visible across the globe. Over the last decade, conversations focused on documenting, anticipating, and preparing for climate risks have provided significant opportunities for interdisciplinary research and for transdisciplinary community building among scientists and practitioners. While some of these opportunities have become visible to contributors to large-scale, interdisciplinary assessments such as the periodic reports issued by the Intergovernmental Panel on Climate Change (IPCC) they are increasingly evident in national- or smaller-scale assessment efforts as have been conducted in the UK, Australia, Canada, the European Union, and in the United States (US).

The Third US National Climate Assessment (NCA3) report (https://nca2014.globalchange.gov) (Melillo et al. 2014) has garnered international attention due to multiple innovations in both process and products. This Special Issue brings together key lessons learned from the NCA3, not only to inform future US assessment efforts, but also to discuss frankly and share broadly what was done, how it was done, what worked and what did not. Our hope and intention behind pulling these lessons together is that those sponsoring, designing, and assisting in assessments at the regional, national and international levels can benefit from

This article is part of a special issue on "The National Climate Assessment: Innovations in Science and Engagement" edited by Katharine Jacobs, Susanne Moser, and James Buizer.

✉ Katharine L. Jacobs
jacobsk@email.arizona.edu

1 Center for Climate Adaptation Science and Solutions, Soil, Water and Environmental Science, University of Arizona, Tucson, AZ, USA

2 Climate Adaptation and International Development, Institute of the Environment, School of Natural Resources and the Environment, University of Arizona, Tucson, AZ, USA

3 Susanne Moser Research & Consulting, Stanford University, Santa Cruz, CA, USA

4 Woods Institute for the Environment, Stanford University, Palo Alto, CA, USA

this experience. Importantly, these articles do not summarize the findings of the NCA3 report itself, but move beyond them to provide insights about the assessment process and outcomes.

2 Background on US national climate assessments

Assessments can be useful at multiple scales, from resolving specific scientific issues to broadly integrating a wide range of sources of knowledge. Climate assessments often include consideration of underlying social, economic, and environmental systems as well as projections of trends in climate-related drivers in complex systems. However, in the case of US national assessments, they also serve as the basis for regulation, policy, and decisions about how to manage risks, which means that they must be conducted with extreme care in order to avoid costly errors.

The 1990 Global Change Research Act established the U.S. Global Change Research Program (USGCRP) and included a requirement that a global change assessment be completed at least every four years that integrates, evaluates, and assesses the state of knowledge of current and projected future impacts.[1] Despite the "at least every 4-years" requirement, only two National Climate Assessments[2] were conducted between 1990 and 2009. There are a variety of reasons why these reports were not completed in a more timely manner, but an important one is that a great deal of infrastructure and social capital is required to conduct assessments properly, given the need to engage stakeholders and external experts in order to meet legal requirements. The federal government does not have the capacity to assess current and projected climate impacts within all of the required sectors without the assistance of external participants. Nor would an assessment conducted entirely within the federal government be as readily acceptable or useful to stakeholders across the US. Importantly, USGCRP has not historically maintained a staff to support timely completion of assessment reports. Rather, new infrastructure and capacity for conducting assessments have been built up each time to support each of the three NCA efforts (for more detail, see Buizer et al. 2015, this issue).

Multiple other large-scale international assessments of research on the implications of global environmental changes have been conducted over the last decades, including the Arctic Assessment (Arctic Council 2005), the Millennium Ecosystem Assessment (2005), and the Ozone Assessments (World Meteorological Organization 2010). In 2007, the National Research Council issued a report that evaluated the lessons learned across this wide array of assessment activities. The findings of this report were highly influential in the development of

[1] Text of the GCRA (1990), **Section 106. SCIENTIFIC ASSESSMENT:** On a periodic basis (not less frequently than every 4 years), the Council, through the Committee, shall prepare and submit to the President and the Congress an assessment which–

1. integrates, evaluates, and interprets the findings of the Program and discusses the scientific uncertainties associated with such findings;
2. analyzes the effects of global change on the natural environment, agriculture, energy production and use, land and water resources, transportation, human health and welfare, human social systems, and biological diversity; and
3. analyzes current trends in global change, both human- induced and natural, and projects major trends for the subsequent 25 to 100 years.

[2] Though the law refers to these assessments as global change assessments, the USGCRP has chosen to refer to them as National Climate Assessments. However, the context for them is clearly broader than climate.

the strategy for the NCA3; its recommendations were explicitly considered for its process and products and are reproduced in Textbox 1:

Textbox 1: Essential Elements of Effective Assessments

- A clear strategic framing of the assessment process, including a well-articulated mandate, realistic goals consistent with the needs of decision makers, and a detailed implementation plan.
- Adequate funding that is both commensurate with the mandate and effectively managed to ensure an efficient assessment process.
- A balance between the benefits of a particular assessment and the opportunity costs (e.g., commitments of time and effort) to the scientific community.
- A timeline consistent with assessment objectives, the state of the underlying knowledge base, the resources available, and the needs of decision makers.
- Engagement and commitment of interested and affected parties, with a transparent science-policy interface and effective communication throughout the process.
- Strong leadership and an organizational structure in which responsibilities are well articulated.
- Careful design of interdisciplinary efforts to ensure integration, with specific reference to the assessment's purpose, users' needs, and available resources.
- Realistic and credible treatment of uncertainties.
- An independent review process monitored by a balanced panel of review editors.
- Maximizing the benefits of the assessment by developing tools to support use of assessment results in decision making at differing geographic scales and decision levels.
- Use of a nested assessment approach, when appropriate, using analysis of large-scale trends and identification of priority issues as the context for focused, smaller-scale impacts and response assessments at the regional or local level.

Source: NRC (2007). *Analysis of Global Change Assessments: Lessons Learned* (http://books.nap.edu/catalog.php?record_id=11868)

NCA3 was explicitly designed to address some shortcomings of previous assessments as well as to benefit from the lessons learned in National Research Council studies, including the America's Climate Choices series (NRC 2010a, b, c, d, 2011). Some of the more salient criticisms of these previous assessments focused on the limited utility of the material produced for "real-world" contexts and the failure to truly connect with the American public. The NCA3 was also very much influenced by the Obama administration's strong focus on decision-relevance, transparency, and planning for resilience, and by the 2012 Strategic Plan of the US Global Change Research Program (USGCRP 2012). That plan included "Informing Decisions" "Sustain Assessments" and "Communicate and Educate" as pillars of the research program, in addition to "Advance Science;" the latter had traditionally been the primary focus of USGCRP's 13-agency science effort. This represented a major shift in policy toward "actionable" science in addition to fundamental climate science research conducted under the auspices of the USGCRP and established a firm foundation for the NCA3's engagement strategy (see, Cloyd et al. 2015, this issue).

Another criticism of previous climate assessments was the burden placed on the scientific community from major efforts, such as the every-six-year IPCC assessment reports and previous US national assessments. The thousands of scientists and other experts who contribute to IPCC and the NCA work as volunteers, with a number of important repercussions. This dependence on volunteers affects the potential pool of available participants and the enthusiasm of the people involved, because it means that many of the same people are tapped for these processes time after time (NRC 2007). The lack of funding and other support for assessment activities continues to plague the NCA assessment process, and was a serious issue within the NCA3. The use of professional NCA staff to provide a much higher level of support for author teams in NCA3 successfully alleviated some (but certainly not all) concerns about the burden on authors. However, through a more strategic and sustained effort, the burden on the scientific community could be further reduced (Buizer et al. 2015, this issue).

3 Overview of contributions to this special issue

This special issue explores the NCA3 as compared with previous assessments from both process and content perspectives. Among the NCA3's important contributions are its emphasis on interdisciplinary learning and the introduction of an engagement strategy that brought hundreds of public and private sector contributors and stakeholders into the assessment community. The NCA3 also included an explicit focus on building sustained assessment capacity, an adaptive approach to managing assessments, and analyzing both the impacts of climate on cross-sectoral systems and the intertwined and cascading effects across sectors. Many new lessons were learned within these efforts and they are discussed in detail in this issue. The first set of articles in this issue describe these NCA3 innovations in some detail. *Building Community, Credibility and Knowledge: the Third US National Climate Assessment* (Jacobs and Buizer 2015, this issue), includes a broad discussion of the role of assessments in general and the NCA3 in particular in framing our understanding of change. The next two papers delve further into process innovations.

Building a Sustained Assessment Process (Buizer et al. 2015, this issue) discusses the rationale for building the infrastructure and capacity for ongoing assessment activities that support a wide range of research and application goals. It provides insights beyond those included in a special report on this topic that was delivered to the USGCRP by the federal advisory committee for the NCA3: *Preparing the Nation for Change: Building a Sustained National Climate Assessment Process* (Buizer et al. 2013).

Engagement in the Third U.S. National Climate Assessment, by Cloyd et al. (2015, this issue) describes both the motivation and the approach used in the NCA3 to build a broad assessment community of scientists, contributors, and stakeholders in regions and sectors across the U.S. Partly because of the broad interest by the public, private, and non-governmental sectors in the activities and conclusions of the assessment, and partly because of the high degree of scrutiny of the process, the entire NCA3 effort was built in the context of balancing the interests of multiple kinds of decision-makers, scientists, and government agencies. A key goal of these engagement efforts was to develop active partnerships that could bring relevant information to the assessment and communicate its findings to audiences and decision-makers across sectors and regions. The authors argue that these partnerships are an essential part of building a sustained assessment process. As many assessment leaders and observers have learned over the past two decades or more, in order for assessment findings to be truly useful from a decision-maker's perspective, the decision-makers themselves need to be part of the process.

The second set of papers focuses on innovations in the assessment process itself that were intended to build assessment capacity over time. To do so, considerable investments were made in producing consistent climate histories and future projections for each of eight US regions and a national set of sea-level-rise projections to allow for comparisons across the nation and to integrate regions and sectors within a common "risk management" framework. Kunkel and colleagues, in their paper, *Innovations in Science and Scenarios for Assessment* (2015, this issue), describe the intent, process, and challenges in doing so. To make such climate histories and projections accessible for impact assessors as well as for downstream users of such information, a significant effort was undertaken during the NCA3 to invest in data management and accessibility, as described by Waple et al. (2016, this issue) in *Innovations in Information Management and Access for Assessments.* This article describes the ongoing efforts of the federal agencies to build a global change information system and to provide transparent access to the data behind each of the major conclusions of the NCA3. The third comprehensive investment in sustained assessment capacity by federal agencies was the development (still ongoing) of an integrated set of national indicators of change across social, physical, and ecological systems and of adaptive responses as described by Kenney and colleagues in *Building an Integrated National Climate Indicators System* (2016, this issue).

The NCA3 also stands out for its considerable effort to integrate across sectors, disciplines, practitioner perspectives and different forms of knowledge systems. Various illustrations of this approach to assessing risk and the status of adaptive responses are provided in the next three articles: first, assessing ecosystem impacts and services in *Climate Change Impacts on Ecosystems and Ecosystem Services in the United States* (Grimm et al. 2015, this issue); the implications of climate change on indigenous peoples and their lands and traditional cultural resources in *Engagement with Indigenous Peoples and Honoring Traditional Knowledge Systems* (Maldonado et al. 2015, this issue); and the integrated social, physical and ecological implications of climate change in coastal areas in *The Third National Climate Assessment's Coastal Chapter* (Moser and Davidson 2015, this issue). Each author team in this set pursued a different path to the goal of providing new insights on the complexities of climate change impacts in the "real world." The collective lessons learned are useful in putting together future assessment teams, designing and supporting assessments, and for connecting these assessments to adaptation processes both within the US and internationally.

Specifically on the question of how assessments should be framed to elevate their decision-relevance and increase their ability to support decisions, Moss (2015, this issue) describes in *Assessing Decision Support Systems and Levels of Confidence to Narrow the Climate Information "Usability Gap"* how important, and yet how difficult it was for NCA3 to improve on past approaches and conventions for assessing scientific confidence and uncertainty. The article makes a strong case for why sustained assessment capacity needs to be built, and why learning from ongoing decision-support efforts, successes, and failures, must be an integral part in improving assessments over time. The article also reviews challenges and approaches for characterizing uncertainties and communicating confidence of lead authors in findings based on the best available—but still incomplete—scientific evidence.

A critical reason for ongoing assessment of the state of knowledge is that both the climate and our understanding of the mechanisms of change are evolving. It is important to evaluate how assessments can support adaptation and adaptive management by taking the perspective of decision-makers who are working toward more resilient systems. In "*Innovations in Assessment and Adaptation,*" Howden and Jacobs (2015, this issue) explore different aspects of the adaptation process and their respective information needs, and suggest some paths forward in building future assessments that address particularly challenging aspects of adaptation.

Finally, the Liverman article (2015, this issue), *U.S. National Climate Assessment Gaps and Research Needs,* discusses critical areas for improving the underlying science foundation for future assessments. The paper does not delineate the typical wish list of research needs from across all common climate science subfields. Rather, it brings into focus two particularly important gaps in knowledge that both limit the understanding of key future vulnerabilities of the US and could undermine adequate preparedness efforts. They are: (a) the need to identify and characterize international linkages that can either amplify or attenuate locally experienced climate risks, and (b) the significant lack of understanding of climate change impacts on the biggest sectors of the economy (such as manufacturing or service industries), as opposed to focusing only on some of the most climate-sensitive sectors (such as agriculture).

In the concluding paper for this special issue, *Aspirations and Common Tensions: Larger Lessons from the US National Climate Assessment,* Moser et al. (2015, this issue) synthesize high-level, integrative lessons from the NCA3, based on the more specific messages and recommendations outlined in each of the preceding papers. We hope these are of wider interest to future US assessment designers and participants and to the international assessment community. They focus on the key ingredients of assessments, including the process, the supporting institutional infrastructure and resource base, the scientific information and foundational data, as well as the people who carry out assessments. These reflections provide frank and detailed insights into the making of the NCA3. Clearly, many of its innovations were improvements over past approaches, building on the extensive national and international experience of its participants. But the NCA3 effort should be viewed as a benchmark and a learning experience to be further improved on in future assessments. The concluding paper points the way to further improvements and invites other assessment designers and leaders to share their experiences for collective learning.

Perhaps the most important message emanating from all the papers in this Special Issue, however, is the importance of not losing momentum in a national climate assessment process that is intended to be sustained, partly because climate risks are evolving so rapidly and partly because the concurrent information needs of users at all levels of government and beyond are also expanding over time. Our perspective is that ongoing assessment processes would be advisable elsewhere in the world as well, and thus it is important that this first "experiment" in building assessment capacity continues to make contributions to climate resilience both nationally and globally over time.

Acknowledgments The Guest Editors for this Special Issue are very grateful to the Editors of Climatic Change, Gary Yohe and Michael Oppenheimer, for their support during this process and for their particular interest in our subjects. We thank Kristin Kuntz-Duriseti and Simon Donner for working through each one of these articles with us and answering our endless questions. We are especially grateful for the help of Mary Black of the University of Arizona/Center for Climate Adaptation Science and Solutions for her expert editing support and to the reviewers for the hours spent providing thoughtful comments. Finally, but most importantly, we are so grateful to all of the authors of these papers, who generously shared their time and insights in developing this collection. Most of them were also intimately involved in the rigors of developing the Third National Climate Assessment, and we are glad that despite those adventures they were still willing to work with us on this Special Issue to harvest our collective lessons learned.

References

Arctic Council (2005) Arctic climate impact assessment—scientific report. Cambridge University Press, Cambridge, 1046 pp

Buizer J, Fleming P, Hays SL, Dow K, Field CB, Gustafson D, Luers A, Moss RH (2013) Preparing the nation for change: building a sustained National Climate Assessment process. National Climate Assessment and Development Committee, US Global Change Research Program.

Buizer JL, Dow K, Black ME, Jacobs KL, Waple A, Moss RH, Moser S, Luers A, Gustafson DI, Richmond TC, Hays SL, Field CB (2015) Building a sustained climate assessment process. Clim Chang. doi:10.1007/s10584-015-1501-4

Cloyd E, Moser SC, Maibach E, Maldonado J, Chen T (2015) Engagement in the third U.S. national climate assessment: commitment, capacity, and communication for impact. Clim Chang. doi:10.1007/s10584-015-1568-y

Grimm NB, Groffman P, Staudinger M, Tallis H (2015) Climate change impacts on ecosystems and ecosystem services in the United States: process and prospects for sustained assessment. Climatic Change. doi:10.1007/s10584-015-1547-3

Howden M, Jacobs K (2015) Innovations in assessment and adaptation: building on the US national climate assessment. Climatic Change. doi:10.1007/s10584-015-1519-7

Jacobs KL, Buizer JL (2015) Building community, credibility and knowledge: the third US national climate assessment. Clim Chang. doi:10.1007/s10584-015-1445-8

Kenney MA, Janetos AC, Lough G (2016) Building an integrated U.S. national climate indicators system. Clim Chang. doi:10.1007/s10584-016-1609-1

Kunkel KE, Moss R, Parris A (2015) Innovations in science and scenarios for assessment. Clim Chang. doi:10.1007/s10584-015-1494-z

Liverman D (2015) U.S. national climate assessment gaps and research needs: overview, the economy and the international context. Clim Chang. doi:10.1007/s10584-015-1464-5

Maldonado J, Bull Bennett TM, Chief K, Cochran P, Cozzetto K, Gough B, Hiza Redsteer M, Lynn K, Maynard N, Voggesser G (2015) Engagement with indigenous peoples and honoring traditional knowledge systems. Clim Chang. doi:10.1007/s10584-015-1535-7

Melillo JM, Richmond TC, Yohe GW (eds) (2014) Climate change impacts in the United States: the third national climate assessment. U.S. Global Change Research Program, 841 pp. (NCA 3 report) https://nca2014.globalchange.gov

Millennium Ecosystem Assessment (2005) Ecosystems and human well-being. Island Press, Washington, DC

Moser SC, Davidson MA (2015) The third national climate assessment's coastal chapter: the making of an integrated assessment. Clim Chang. doi:10.1007/s10584-015-1512-1

Moser SC, Melillo JM, Jacobs KL, Moss RH, Buizer JL (2015) Aspirations and common tensions: larger lessons from the third US national climate assessment. Climatic Change. doi:10.1007/s10584-015-1530-z

Moss RH (2015) Assessing decision support systems and levels of confidence to narrow the climate information "usability gap". Clim Chang. doi:10.1007/s10584-015-1549-1

National Research Council (2007) Analysis of global change assessments: lessons learned. National Academies Press, Washington, D.C., 198 pp

National Research Council (2010a) America's climate choices: adapting the impacts of future climate change. National Academies Press, Washington, D.C., 292 pp

National Research Council (2010b) America's climate choices: advancing the science of climate change. National Academies Press, Washington, D.C., 528 pp

National Research Council (2010c) America's climate choices: informing an effective response to climate change. National Academies Press, Washington, D.C., 276 pp

National Research Council (2010d) America's climate choices: limiting the magnitude of future climate change. National Academies Press, Washington, D.C., 276 pp

National Research Council (2011) America's climate choices. National Academies Press, Washington, D.C

U.S. Global Change Research Program (2012) The National Global Change Research Plan 2012–2021: a strategic plan for the U.S. Global Change Research Program. USGCRP Report. http://data.globalchange.gov/assets/d0/67/6042585bce196769357e6501a78c/usgcrp-strategic-plan-2012.pdf

Waple AM, Champion SM, Kunkel KE, Tilmes C (2016) Innovations in information management and access for assessments. Clim Chang. doi:10.1007/s10584-015-1588-7

World Meteorological Organization (2010) Scientific assessment of ozone depletion. World Meteorological Organization, Global Ozone Research and Monitoring Project, Report no. 52. http://www.esrl.noaa.gov/csd/assessments/ozone/2010/report.htm

Climatic Change (2016) 135:9–22
DOI 10.1007/s10584-015-1445-8

Building community, credibility and knowledge: the third US National Climate Assessment

Katharine L. Jacobs[1] **· James L. Buizer**[1]

Received: 16 February 2015 / Accepted: 8 June 2015 / Published online: 11 July 2015

Abstract Assessments that are designed to be credible and useful in the eyes of potential users must rigorously evaluate the state of knowledge but also address the practical considerations—politics, economics, institutions, and procedures—that affect real-world decision processes. The Third US National Climate Assessment (NCA3) authors integrated a vast array of sources of scientific information to understand what natural, physical and social systems are most at risk from climate change. They were challenged to explore some of the potentially substantial sources of risk that occur at the intersections of social, economic, biological, and physical systems. In addition, they worked to build bridges to other ways of knowing and other sources of knowledge, including intuitive, traditional, cultural, and spiritual knowledge. For the NCA3, inclusion of a broad array of people with on-the-ground experience in various communities, sectors and regions helped in identifying issues of practical importance. The NCA3 was more than a climate assessment; it was also an experiment in testing theories of coproduction of knowledge. A deliberate focus on the assessment process as well as the products yielded important outcomes. For example, encouraging partnerships and engagement with existing networks increased learning and made the idea of a sustained assessment more realistic. The commitment to building an assessment focused on mutual learning, transparency, and engagement contributed to the credibility and legitimacy of the product, and the saliency of its contents.

1 Introduction: learning from assessments

Global change assessments by definition must incorporate both social and physical science and consider the potential implications of current and future sources of stress and opportunity

The authors are members of the development team for the Third National Climate Assessment, specifically, the Director (Jacobs) and a member of the Executive Secretariat (Buizer), and thus are not unbiased observers.

This article is part of a special issue on "The National Climate Assessment: Innovations in Science and Engagement" edited by Katharine Jacobs, Susanne Moser, and James Buizer.

✉ Katharine L. Jacobs
jacobsk@email.arizona.edu

1 University of Arizona, Tucson, AZ, USA

(NRC 2007). The Third US National Climate Assessment (NCA3, Melillo et al. 2014) provided participants an opportunity for learning, not only about how to integrate a wide array of information, but also how to knit together different ways of knowing, and how to make assessments more relevant to decision-makers. This paper describes how this assessment—building on lessons from other US and international assessments—was designed, and the processes through which it was implemented. It delineates key elements of success and describes some challenges encountered by those involved, aiming to inform future assessment processes in the US and elsewhere.

Despite a vast array of available sources of scientific information, from satellite data and ocean observing systems, to information about human and ecosystem vulnerabilities, to social science and economics data, understanding what is most at risk from climate change is a daunting task. When these sources of data are integrated with knowledge held by on-the-ground resource managers, business executives, or policy-makers, and a range of geographic and time scales are represented, there is a virtual explosion of data sources, ways of understanding the world, and potential paths for analysis and interpretation. This challenged NCA3 authors to step outside of more familiar, traditional, and disciplinary approaches to explore some of the potentially substantial sources of risk that occur at the intersections of social, economic, biological, and physical systems. They also had to make difficult choices about which issues to emphasize, given the limits of a "manageable" final report. With initial instructions to most teams to condense their work into eight pages, only the most critical issues could be included, even if underlying foundation reports were hundreds or in some cases thousands of pages. This led to tradeoffs about perceptions of risk and personal values within each author team, and significant opportunities for learning in an interdisciplinary and transdisciplinary context.

Assessments that are designed to be credible and useful in the eyes of potential users must rigorously evaluate the state of knowledge but also address the practical considerations—politics, economics, institutions, and procedures—that affect real-world decision processes. In addition, they often must build bridges to other ways of knowing and other sources of knowledge, including intuitive, traditional, cultural, and spiritual knowledge. These need to be respected and incorporated if assessments are to be useful and meaningful to multiple audiences, including Native Americans and a range of other communities.

2 Background on the Third National Climate Assessment process

The National Climate Assessment responds to the 1990 Global Change Research Act (GCRA), Section 106, which requires that an assessment be prepared at least every 4 years that synthesizes, analyzes, evaluates, and assesses the knowledge developed within the US Global Change Research Program (USGCRP), but also identifies impacts across a series of sectors and projected timeframes. Only three such integrated assessments have been completed in the 24 years since 1990. The USGCRP, which was created by the GCRA, is a consortium of thirteen federal science agencies that is charged with coordinating and managing the substantial federal investments in climate science (https://globalchange.gov). Though these assessments are required every 4 years, only two previous assessments were completed (in 2000 and 2009), for a variety of reasons discussed elsewhere (see Buizer et al. 2013).

The first two National Climate Assessments focused primarily on summarizing outcomes of research and documenting the state of knowledge, generally focusing on a synthesis of published, peer-reviewed literature. However, they varied significantly in the extent of stakeholder involvement. For a number of reasons (see Buizer et al. 2013) their outcomes were not as widely used as the authors had hoped. The NCA3 builds on these assessments, but also reflects a new sense of urgency from scientists, managers, and decision-makers across the country—based on growing evidence that climate change is no longer only an issue for the future, but a problem that requires action now. Because impacts are now observable in every region and sector across the country, the products and processes of this assessment were more explicitly designed to be useful in decision contexts. The increased visibility of impacts changed the tenor of the NCA3 conversation and encouraged a stronger focus on assessing response strategies than in previous assessments.

Individuals, communities and companies across the globe are now focusing on building a more resilient future. The terms they use, such as *sustainability, adaptation, resilience,* and *preparedness,* imply a wide variety of activities, but those working in these arenas have a great deal in common. They may work in their communities to reduce use of fossil fuels, promote health care for the urban poor, or manage invasive species in a wildlife refuge, but all are concerned about the health of the planet and of human and natural systems in the face of disruption and change. The NCA3 brought together people with diverse interests and expertise; some participants had engaged in previous assessments, others were specifically included to bring new capacity and new ideas into the process. A deliberate focus on community-building within the NCA3 allowed new relationships to be forged among those who otherwise might never have met. The investments in relationship-building within the assessment recognized the interdisciplinary nature of the scientific challenges, but was also intended help sustain the assessment process beyond the report and improve the relevance of the final products in connecting science to decision-making. Consequently, this assessment was "owned" by a much larger group of individuals because of the NCA network of partner organizations and the outreach efforts through NGOs and professional societies (Cloyd et al., Submitted for publication in this special issue).

3 Building the NCA3 process

Initial partners in the NCA3, especially federal agency representatives, staff in the US Global Change Research Program Coordination Office, and the NCA3 federal advisory committee, created a common mission and vision statement based on the perceived need to support decisions and be inclusive and useful to a broad audience of decision-makers and citizens across the US and the globe (USGCRP 2011). They quite consciously worked to create and test innovative and collaborative products and processes that would be useful and possible at this time in US history, recognizing the growing need for "actionable information" to support adaptation and mitigation.

Building the capacity for this type of assessment meant involving people who know how to remove the barriers between scientific disciplines and those who can build connections with user communities in regions and sectors that may use very different language and terminology. Facilitators, communicators, and "boundary spanners" (those who help connect scientists and scientific information with decision-makers and the public) (Guston 2001; Carr and Wilkinson 2005; Hoppe 2010; Lemos et al. 2012) were deliberately incorporated in the NCA3 process because of the need to bridge the gap between scientists and decision-makers in public and private organizations, government agencies, and businesses. The lessons learned over the last decade about the importance of managing the boundary between science and decision-making, e.g., in the context of the National Oceanographic and Atmospheric Administration's (NOAA)

Regional Integrated Science Assessments (RISAs), and many others were explicitly incorporated into the design of the NCA3. In that sense, the NCA3 can be viewed as a large-scale experiment designed to test theories about knowledge networks (Jacobs et al. 2010; Bidwell et al. 2013) and knowledge-to-action research.

The authors, federal advisory committee members, and staff of the NCA3 felt that building the interdisciplinary and regional capacity to *sustain* this effort would be critical to ensuring that its findings are useful in decision-making. Stakeholders who participated in NCA3 town halls across the country indicated that they were more likely to use the results of the NCA3 if they knew that they could count on the NCA process to produce rigorous and relevant information over time. From their perspective, relevance of the information produced was directly related to their own ability to access, understand, and contribute to the process. Similarly, the importance of building relationships among scientists and stakeholders during the assessment process was recognized as essential in ensuring credible outcomes that are also perceived as relevant and usable. More than a report based on peer-reviewed literature, the inclusive process itself was viewed from the beginning as an important outcome of the NCA3 effort.

4 The National Climate Assessment Development and Advisory Committee (NCADAC): lessons learned

In part because of the decision to make the NCA3 assessment more inclusive and transparent than previous national assessments, the federal advisory committee that took responsibility for producing the NCA3, the National Climate Assessment Development and Advisory Committee (NCADAC) was unusually large, including 44 non-federal participants and 16 federal agency ex-officio representatives. Establishing this committee took 18 months – far more than the original 6—month estimate. This significantly impacted the overall schedule for the assessment and reduced the time available to get the report done: rather than the 4 years allotted by statute to complete the assessment, the NCADAC essentially had only 2.5 years. These regulatory and political transaction costs need to be understood and factored into future assessment planning to minimize barriers to progress and optimize outputs and outcomes from ongoing assessment processes.

Although establishing the NCADAC caused a major delay in the report development, the time in the "holding pattern" was well spent. Because the advisory committee was not yet in place, no decisions could be made—so a series of methodology workshops were set up as "listening sessions" to help inform the process. These NCA methodology workshops (NCA report series, https://globalchange.gov) established a foundation of common knowledge among participants and built capacity for subsequent assessment activities. They gave the participants time to assess priorities, solidify goals and objectives of the NCA3 effort, build knowledge of assessment processes, and broaden the participant community so there could be a collective path forward. In fact, separate communities started to evolve around the workshop topics, including information management, communications, indicators of change, scenarios and modeling, vulnerability assessment, and valuation techniques. Most visible and active of these communities today is the group that focused on indicators; it has developed a broad vision and a pilot demonstration within USGCRP for social, physical, and environmental indicators of climate change (Kenney and Janetos, Submitted for publication in this special issue).

Based on its size, there were fears that the NCADAC would be unwieldy and expensive to support, but surprisingly few major problems arose. This was in part because of strong leadership, which included three experienced chairs and a 12-member Executive Secretariat. The latter included individuals with significant experience in previous assessments who had a wide range of disciplinary, legal, and engagement expertise. Executive Secretariat input was solicited on all process and content issues prior to presentation of these ideas for review and decision by the broader committee. Inclusion of a range of people on the NCADAC who were process experts was another unusual aspect of this advisory committee that served the overall assessment extremely well. Because webinars and conference calls were frequently used instead of in-person meetings and the in-person meetings themselves could be generously characterized as "frugal," the costs of supporting the NCADAC were far lower than anticipated.

Another critical decision was to include the USGCRP agencies, the chair of the USGCRP, the White House Council on Environmental Quality (CEQ), and the Department of Homeland Security as full participants in the NCADAC, albeit as non-voting members.[1] Given that the full NCADAC agreed to make decisions by consensus rather than by vote, the limitation on voting by the 16 federal members was not as significant as it might otherwise have been. But having the federal agencies engaged as equal participants and encouraging them to make their scientific and other resources available to support the process was critical to achieving approval of the final document. Although they could not participate in the consensus decision to approve the final draft report, government scientists played important roles in providing scientific expertise on multiple topics and were active in author teams. This balance worked in favor of a more credible and defensible product. Without government input in the process of developing the draft NCA3 report, it is highly unlikely that a consensus could have been reached (or as easily reached) on the ultimate product. Without such a consensus, the document might not have been accepted and released as a government document.

5 Leadership

Not surprisingly, the NCA3 products and process reflected the values and experience of those who led it and of the 30 author teams[2] who were selected by the NCADAC. Running a process that had so much visibility and such high expectations can be a daunting task even with a small group of participants; the involvement of 60 people on the NCADAC and of hundreds to thousands more in authoring underlying documents raised the stakes substantially. As with both previous US assessments, navigating the scientific and policy issues that arose almost daily over a two-and-a-half-year period was a major challenge. The assignment was to develop an unassailable scientific document through a transparent and inclusive process while avoiding potential political pitfalls and practical irrelevance. Enlisting authors and reviewers who were as representative as possible of their respective expert or stakeholder communities and who had impeccable credentials was critical to the ultimate success of the endeavor. Balancing many different interests and scientific disciplines required a delicate hand and chapter authors who were willing to step outside of their comfort zone to experiment with new ways of learning and knowing.

[1] The Council on Environmental Quality provides oversight on regulatory and policy matters related to natural resources and the environment. It is parallel to the Office of Science and Technology Policy within the White House. The Department of Homeland Security—which includes the Federal Emergency Management Agency—did not exist at the time of the formation of the USGCRP in 1990, but is now very engaged in climate-related matters and chose to join the NCADAC as a non-USGCRP agency.

[2] Each chapter was led by two coordinating lead authors and typically had 6 additional authors, resulting in a total of approximately 240 primary plus ~60 contributing authors of the whole report.

The NCADAC was led by a trio of very experienced leaders. The chair, Dr. Jerry Melillo of the Marine Biological Laboratory at Woods Hole, co-led the two previous NCAs and has had a distinguished career as an ecologist spanning decades of work on climate-related topics. As an economist,, co-chair, Dr. Gary Yohe (Wesleyan University), provided important insight on costs and consequences of climate impacts, while also providing important linkages to the IPCC Fifth Assessment. Terese (T.C.) Richmond, the second co-chair and a natural resources lawyer from Seattle, represented stakeholder and private sector interests on the leadership team and worked to ensure that outcomes were useful for decision-makers. In many ways she took on the role of an ombudsman, ensuring that concerns of individual NCADAC members, authors, and staff were properly addressed.

Another element that contributed to meeting the strategic goals of the assessment was the willingness of the leadership to argue strongly for positions, yet compromise for the good of the process at the right moment. Given the ambitious expectations, there had to be trust among participants and an awareness that the collective outcome was more important than winning personal battles. Matching the capacity of the leadership to the nature of the challenge is important to the success of assessments, especially as the nature of these challenges changes over time.

The leadership also had to balance the ambitious goals of the NCADAC, author teams, and staff with what was actually "doable" from the perspective of the authors and the federal agencies. For example, NCADAC members suggested that each author team should have expert observers/assessment specialists who could assist with the consistent characterization of uncertainties, but this was not possible due to a lack of time and resources. Many decisions made during the assessment represented a compromise between an optimal approach and what was possible, balancing scheduling constraints with concerns about quality and accuracy.

6 Sources of knowledge

NCA3 took advantage of multiple sources of knowledge, ranging from traditional ecological knowledge of Indigenous Nations to the latest satellite technology. Recent advancements in understanding climate communication, public perceptions, and information systems were incorporated as well. Many participants noted the richness of the conversation that took place within author teams, due at least in part to the transdisciplinary nature of the assignment. Explicitly focusing on sources of risk and topics of greatest concern within regions and sectors, as opposed to starting with climate drivers, was very helpful in reframing conversations in ways that were more meaningful to decision-makers. A team of communication experts provided advice on a wide range of issues, not the least of which were how to explain complex issues simply and how to focus on communication outcomes that could reach a wide range of audiences. For example, there was an effort to make sure the graphics and associated captions painted a coherent picture that reinforced the text across the whole report, because many people learn visually or by examples rather than through reading text carefully.

A benefit of the large advisory committee was its topical diversity, providing the NCA3 with subject expertise and sources of information not only on climate science and regional and economic sector impacts but also other important topics and issues, such as decision support, adaptation, and mitigation. The wide range of strategies, participants, and contributors, along with the highly transparent approach to conducting this assessment, probably contributed to the overall credibility of the final report and its broad appeal across the US. Further, the strong emphasis on diversity on the NCADAC and the inclusivity of the engagement strategy may have contributed to the absence of legal challenges to date.

7 Documenting scientific findings and levels of certainty: traceable accounts

Another way to understand how different sources of knowledge were integrated into individual chapters of the NCA3 is through the traceable accounts, an NCA3 innovation designed to document the process that the authors used to reach their conclusions and describe their level of certainty. Authors were asked to go beyond standard referencing conventions that documented their scientific sources and describe how they selected the key issues, which literature they depended on most, and which scientific uncertainties are most important now and in the future. This highly transparent approach enhanced the clarity of the process and avoided heavy reliance on terms like "likely" and "virtually certain" used in the text of other assessments to characterize certainty in ways that most audiences either do not understand or interpret in widely different ways (Ekwurzel et al. 2011; Morgan et al. 2009). It will be interesting to see if subsequent assessment processes take advantage of this approach.

Inclusion of direct links to data used to support conclusions and to the references for each climate science graphic reinforced the robust nature of the report's conclusions. Documenting both the thought processes and the data used built the credibility of the NCA3 and provided information of interest to more sophisticated users. Though hard to evaluate, it would be good to know whether potential critics of the NCA3 process accessed the data and found it convincing. The transparent documentation avoided the accusation made in past assessments, however, that the science was a "black box."

8 Coproduction and assessment

The term "coproduction of knowledge" is useful in describing mutual learning between scientists and stakeholders (Lemos and Morehouse 2005; Mauser et al. 2013). The theory behind it—that if a scientific product is intended to be used by decision-makers, the decision-makers need to be involved in the problem definition, the discussion of solutions, and creation of the product—was definitely supported in the context of this assessment. A primary motivation for collaborative knowledge production in an applied context is to facilitate access to the facts for decision-makers and to help scientists understand how that information is used. Identifying what motivates stakeholders to feel ownership and see value in the information is a challenge, but one way to enhance both the perception and the reality of relevance to decision processes is to engage them in generating that information. Experience in NOAA RISAs (e.g., Dilling and Lemos 2011; Lemos et al. 2014; McNie 2008) has shown that scientists tend to be viewed as more legitimate sources of information if they actively engage in long-term relationships with stakeholders, build their own understanding of the tacit knowledge of practitioners, and come to be trusted by them.

The concept of coproduction as an approach to engagement was reinforced by experience in the NCA3 effort: the NCA3 report itself served as a convening point for conversations between stakeholders and scientists. They negotiated over which topics to cover, what the evidence was, and where the remaining uncertainties were. As a result, communities formed that can be the core of a sustained assessment process (Buizer et al. 2013). An example of the community-building effort was the facilitated engagement of author teams in sorting through multiple technical input documents and peer-reviewed literature in order to agree on chapter key messages and a process for writing the supporting material (e.g., Moser and Davidson, Submitted for publication in this special issue).

Embedded in this concept of coproduction is the perception that how you engage people in assessment and who to engage in assessment processes is at least as important as the findings themselves. Strategic engagement with the specific stakeholders who are most likely to benefit from a conversation with scientists is an important place to start, but training scientists in how to be receptive to tacit knowledge of on-the-ground experts is at least as important. The coproduction approach should optimize the experience of the participants while also working toward the desired end-point of the research itself. However, the goal of achieving long-term trusted relationships in the process of assessment is not always attainable; not all researchers are prepared to invest the time and energy required to engage in useful ways with decision-makers and vice-versa.

Based on personal communications with members of the NCADAC, an interesting outcome of the NCA3 process was that the more participants were asked for their input about the assessment *process*, the more invested they became in it. For example, some NCADAC members who were initially skeptical about whether the process was workable became much more interested and supportive as they saw their own ideas bearing fruit, learned to value the input of others, and in some cases became more willing to dedicate more time and give input on options for the path forward. The early methodology workshops also fostered a more thoughtful and inclusive process. This primed the pump for the deliberate and ambitious coproduction process that subsequently developed.

The concepts of use-inspired research and decision support influenced the structure, framework, and components of the NCA3 report, including the selection of chapter topics and the directions to authors to focus on what was most at risk. For example, the explicitly cross-sectoral chapters (Energy, Water and Land; Urban Systems, Infrastructure, and Vulnerability; Biogeochemical Cycles; Rural Communities; Indigenous Peoples, Lands, and Resources; Land Use and Land Cover Change) illustrate the expansion from the narrow sectoral and disciplinary approach in previous assessments to one that embraced the complexity of real-world challenges. Similarly, the input from users resulted in strong direction to authors to focus on multiple stresses and issues that matter to communities, businesses, and policy-makers. All key characteristics of the report format (strictly limited chapter length, the inclusion of key messages for each chapter, a searchable web format, and use of strong graphics and images illustrating climate change impacts and responses) derived from the strong desire to make it both useful and used. Further, the way the report was delivered through partner networks and trusted intermediaries reflected an understanding of what makes information trusted and useful to stakeholders (Cloyd et al., Submitted for publication in this special issue).

9 Assessment and knowledge networks

A guiding principle in the engagement strategy of the NCA3 was to build capacity for assessment by tapping into the strength of existing professional and private networks (USGCRP 2011). A "network of networks" approach was explicitly adopted, with the intent to rapidly expand outreach and the ability to harvest knowledge from external groups (Cloyd et al., Submitted for publication in this special issue). Knowledge networks, defined here as groups of people who have intersecting interests and who choose to engage with each other (e.g., via the Internet, social media or in person) to share information and knowledge about selected topics, are increasingly seen as a powerful means to address complex problems (Dyer and Hatch 2006; Jacobs et al. 2010; Eden 2011; Bidwell et al. 2013; Kirchhoff 2013). The most deliberate network-building effort of the NCA3 was the NCAnet, which now involves more than 150 organizations representing at least 100,000 members. (For a discussion of their importance in the NCA3 outreach strategy and helping to sustain the process, see Cloyd et al. (Submitted for publication in this special issue)).

The concept of knowledge networks also permeated the development of working groups of the NCADAC, including the Executive Secretariat. Within the overall umbrella of the NCADAC, each Secretariat member was assigned to organize particular activities that occurred within the purview of the NCADAC, including coordination of groups of chapters (e.g., regions, sectors); report review and response processes; engagement and communication; approach to the sectoral analyses; and guidance on characterizing uncertainty and "traceable accounts." All Secretariat members also had assignments to ensure the internal consistency of the information across the 30 chapters of the report. Several members served as convening lead authors of chapter teams and most were authors on at least one chapter. Because they had roles as authors as well as leaders of the NCADAC, they could evaluate whether the teams were following the guidance that had been provided. This tiered approach to authorship and oversight was crucial to effectively managing a very complex system with multiple moving parts. In addition, deliberate community building within and between the author teams, the NCADAC, and the staff, included social events that helped build personal and professional relationships. This was difficult, given government travel restrictions, but a great deal was accomplished within the meetings that did take place.

10 Electronic innovation

From the beginning, key decision-makers for the NCA3 at NOAA (the supporting agency for the Federal Advisory Committee and home of the Technical Support Unit) and at the Office of Science and Technology Policy in the White House favored a "new generation" of assessment that was entirely electronic, because of the rapidly changing information world and the administration's commitment to innovation. The plans for web-based delivery of the NCA3 strongly impacted both the product development and the experience of the authors. This was one of the first major government reports delivered via the Internet, and this mode greatly enhanced both accessibility of the products and traceability of the findings. Based on reactions from government and outside stakeholders, this approach was extremely successful and will influence future assessments.

Web-based delivery provides instant access through search engines on the web, ensuring that the NCA3 report can be regularly accessed by people seeking answers to climate-related questions, not just by those who already know what a National Climate Assessment is. The linked data also allow more sophisticated decision-makers to "look behind the conclusions" and directly access the evidence underlying them. Electronic delivery, including compatibility with Facebook and Twitter, also meant that a broader audience was engaged in the NCA3 rollout process.

The development of an interagency Global Change Information System as the underlying data platform for the NCA3 is intended to be the foundation of a long-term interagency investment of the USGCRP that would allow interoperability and data access for a wide array of potential uses, including support of some "live" indicators of change that could be updated more regularly than the quadrennial reports (Kenney and Janetos, Submitted for publication in this special issue and Waple et al., Submitted for publication in this special issue).

Other electronic platforms were important during the development and review of the assessment report. For example, chapter authors were provided with easily accessed electronic workspaces, which helped to facilitate draft development, author collaboration, and the assessment process itself. However, version control was a critical concern and required negotiated agreements about how authors, editors, and staff could amend chapters, especially in later iterations. Status spreadsheets were used to track responses to each of an estimated 10,000 review comments over the duration of the NCA3 process. This was very time-consuming and tedious, requiring intense central management and coordination. On the other hand, the comment-response process and the technical input reports (from the public) were received and documented through automatic online mechanisms and carefully tracked by the central NCA3 staff; these more automated systems worked extremely well.

11 Sources of tension and lessons learned

The NCA3 is simultaneously a summary of the state of knowledge of a changing climate in the US, a process for engaging researchers and stakeholders, and an assessment that depended on coproduction of knowledge between users and experts. The NCA3 was and is path-breaking in all three respects. But these aims are not necessarily easily reconciled as the reflections below demonstrate.

There were many specific sources of tension, including the need for busy professionals, all volunteering their time, to meet deadlines while developing the highest-quality outcomes possible. For example, despite the collective agreement to prioritize the needs of managers and policymakers for the NCA3 report, there were differences of opinion about how ambitious the report should be in moving beyond GCRA legal requirements to be as supportive as possible of decision-makers' needs for scientific information. Instructions to authors on handling issues such as characterizing risk and levels of certainty were not followed to equal degrees by all of the teams—there were so many complexities in the process and so many guidance documents that some authors moved forward without explicitly following the guidance. Further, there were ongoing concerns about internal consistency across the 840 pages of the final report. A final challenge was ensuring readability and consistency of the document given the influence of 300 individuals in drafting it. In the end, all of these concerns were resolved, resulting in unanimous approval by the NCADAC of the final report. This outcome was primarily due to the hands-on efforts of the NCADAC leadership and NCA3 staff developing a consensus approach to each unresolved issue with the affected authors. Understanding these process-related tensions and preparing for them can minimize the challenges for future assessments.

Use of non-traditional material in the NCA3 report also posed challenges. Although the NCADAC agreed from the beginning that it was important to incorporate experiential knowledge in addition to standard peer-reviewed literature, there were ongoing debates about exactly how to do this. In many cases the quality of the data was indisputable because it derived from very highly reviewed or vetted government sources even if not from official "peer-reviewed" literature. However, even in the context of the NCADAC's publicly and federally approved guidance document on how to handle information quality, there were questions from federal reviewers about several of the sources used. An important innovation in the NCADAC guidance given on information quality was the requirement to make information use consistent with the quality of the data. For example, information used as illustrations and case studies does not require the same kind of academic review that sources of major science conclusions do (USGCRP 2012). Given known challenges in all review processes, including peer-review, the status of peer-reviewed literature relative to other sources is worth discussing carefully. Future assessments will need to address this problem and to identify which data are critical to include even if not peer-reviewed.

There were also lessons learned by physical scientists about handling the evolving understanding of climate drivers and impacts. One issue was how to include new scientific insights that developed over the course of the assessment. The most important debate had to do with handling the new Community Model Intercomparison Project (CMIP) 5 inputs, which were not available early enough in the NCA3 process to be useful for impact assessment. However, the science chapter authors were adamant that this material had to be included if the NCA3 were to be taken seriously (Kunkel et al., Submitted for publication in this special issue). In turn, authors of impacts chapters insisted that studies that did not use the standard climate scenarios used in NCA3 should be eligible for inclusion. The NCA leadership set up special subcommittees to address science issues where there were significant debates so that broadly acceptable outcomes could be negotiated.

Extreme events occurring during the NCA3 assessment process (including Superstorm Sandy) brought up new issues about prediction and physical mechanisms of climate impacts but also highlighted the importance of interdisciplinary work and cascading effects, especially linkages among systems and the vulnerability of urban infrastructure, reinforcing earlier conclusions about the importance of systems thinking and the potential for catastrophic failures.

12 Integration of knowledge and sustained assessment

A number of assessment authors and NCADAC members indicated in personal communications that they were willing to stay engaged throughout the relatively arduous NCA3 process in part because they expected to build the infrastructure and capacity for future assessments, not just create a single report. A principal frustration voiced by participants in previous assessments was that assurances about future engagement with them were not realized. Several participants mentioned that if this were to happen again after the explicit commitments made in the NCA3 process, future engagement with stakeholders as well as scientists would be seriously impaired. Many NCA3 participants hope that the now-trained array of "assessors" across the country will assure that some form of assessment activity will continue even if federal leadership for a sustained assessment does not materialize.

Despite this investment in assessment capacity, some teams were better at facilitating inter- and trans-disciplinary conversations than others. Previous experience in bridging the gap between science and decision-making was one criterion in the selection of authors and members of the NCADAC. Inclusion of industry, government, and NGO representatives in the NCADAC and in chapter teams also helped to ensure that the topics were relevant to decision-making and the degrees of certainty about the findings were clear and defensible. An example of this helpful input from stakeholders is evident in the relatively major changes that evolved in the decision-support chapter (Moss et al. 2014), from a relatively theoretical public draft to a final version that included more examples and conclusions based on managers' experience.

Integration of knowledge across sectors and regions was greatly enhanced in the rigorous review process. Up to thirty versions of some of the chapters were prepared over an 18-month period, each responding to new input from author teams, the NCADAC, the Executive Secretariat, external chapter reviewers, federal agencies, the National Academy of Sciences, the White House, and the public. In some cases, inconsistencies were identified across chapters that led to important interdisciplinary discussions and new ways to clarify, explain, and defend scientific understanding. In others, the conclusion was that more research was required to resolve the issues at hand. This thorough review process resulted in a much more robust report and far less criticism than might otherwise have been expected.

13 Staff contributions

To ensure effective communications across chapter teams and among the many players involved, NCA staff and editors were assigned to author teams and NCADAC working groups to support them in numerous ways, such as meeting planning or facilitation, assistance with graphics and writing, filling important knowledge gaps, coordination across chapters, and audience-tailored translation of scientific text. The fact that the professional staff had significant content knowledge, good writing and communication skills, and authority to engage with the author teams as needed was an important component of success and should be replicated in future assessments.

Central staffing for the NCA3 included experienced professionals (at the USGCRP office and the NOAA Technical Support Unit) with expertise in a wide range of sectoral and scientific topics; their contributions to the overall process were significant. The level of staff support for author teams was much more visible in this process than in other assessments. High-level staff commitment and quality input often helped authors meet key deadlines. Graduate students and others from partner organizations and universities provided additional support to many teams. The effort, motivation, and capacity of staff members was widely noted as exceptional in conversations, in public meetings, and by authors and NCADAC members in the final evaluation questionnaire. In such complex processes there is a need for trusted staff who can meet expectations without biasing the outcomes.

A critical factor in the ultimate success of NCA3 was the careful work of those who edited and prepared the document for electronic delivery, designed the graphics, trained the participants for the deluge of media requests, conducted town halls to share the draft in all of the regions of the US, and checked and rechecked the responses to comments. A notably smooth rollout process in May of 2014 (and beyond) resulted from years of planning and preparation, the help of the NCAnet, and the huge contributions of climate communications professionals and their networks. The many people involved in every aspect of the report development and review meant that there had been a great deal of socializing of the contents prior to the release, ensuring high credibility and very limited criticism of the findings.

14 Conclusions

Learning from assessment processes involves personal and collective experience and judgment; there is a human element of assessments that is not typically addressed in academic literature. For the NCA3, inclusion of a broad array of people who both study climate change and experience it in their personal lives helped in identifying issues that really matter. Many stakeholders viewed building a sustained assessment process as an investment in relevance in a decision context. Other critical ingredients of decision relevance included using interdisciplinary, risk-based approaches, providing transparent access to data and evidence, and framing controversial topics in unbiased ways. Strong leaders and staff who understand what is achievable, who can harness the power of knowledge networks and benefit from technology and electronic innovations contributed to the success of assessment processes and products.

The NCA3 was more than a climate assessment; it was also an experiment in testing theories of coproduction of knowledge. The development of author teams that included both scientists and decision-makers was a central tenet of the assessment approach and allowed integration of multiple kinds and sources of knowledge. Explicitly incorporating coproduction had two benefits: the potential for more insightful information about impacts and increased relevance to decisions. The highly engaged NCA3 staff and leadership also advanced knowledge integration and sharing across the assessment enterprise.

Finally, encouraging the use of existing sectoral, regional, professional, and academic networks as partners (particularly the NCAnet "network of networks") increased learning and made the idea of a sustained assessment more realistic. The commitment to building an assessment focused on mutual learning, transparency, and engagement contributed to the credibility and legitimacy of the product, the saliency of its contents to abroad public, and to the absence (to date) of legal challenges.

References

Bidwell D, Dietz T, Scavia D (2013) Fostering knowledge networks for climate adaptation. Nat Clim Chang 3: 610–611

Buizer JL, Fleming P, Hays SL, Dow K, Field CB, Gustafson D, Luers A, Moss RH (2013) Report on preparing the nation for change: Building a sustained National Climate Assessment process, National Climate Assessment and Development Advisory Committee. http://downloads.globalchange.gov/nca/NCADAC/NCADAC_Sustained_Assessment_Special_Report_Sept2013.pdf

Carr A, Wilkinson R (2005) Beyond participation: boundary organizations as a new space for farmers and scientists to interact. Soc Nat Resour 18(3):255–265

Cloyd E, Moser S, Maibach E, Maldonado J, Chen T (submitted for this issue) Engagement in the Third US National Climate Assessent: Commitment, Capacity, and Communication for Impact. Climatic Change

Dilling L, Lemos MC (2011) Creating usable science: opportunities and constraints for climate knowledge use and their implications for science policy. Global Environ Chang 21(2):680–689. doi:10.1016/j.gloenvcha.2010.11.006

Dyer JH, Hatch NW (2006) Relation-specific capabilities and barriers to knowledge transfers: creating advantage through network relationships. Strateg Manag J 27(8):701–719

Eden S (2011) Lessons on the generation of usable science from an assessment of decision support practices. Environ Sci Policy 14:11–19. doi:10.1016/j.envsci.2010.09.011

Ekwurzel B, Frumhoff PC, McCarthy JJ (2011) Climate uncertainties and their discontents: increasing the impact of assessments on public understanding of climate risks and choices. Clim Chang 108:791–802

Guston DH (2001) Boundary organizations in environmental policy and science: an introduction. Sci Technol Hum Values 26(4):399–408

Hoppe R (2010) Lost in translation? Boundary work in making climate change governable. In: Driessen P, Leroy P, van Vierseen W (eds) From climate change to social change: perspectives on science-policy interfaces. International Books, Utrecht, pp 109–130

Jacobs K, Lebel L, Buizer J, Addams L, Matson P et al (2010) Linking knowledge with action in the pursuit of sustainable water-resources management. Proc Natl Acad Sci. doi:10.1073/pnas.0813125107

Kenney M, Janetos A (submitted for this issue) Building an Integrated National Climate Indicator System. Climatic Change

Kirchhoff C (2013) Understanding and enhancing climate information use in water management. Clim Chang 119(2):495–509

Kunkel K, Moss R, Parris A (submitted for this issue) Innovations in Science and Scenarios for Assessment. Climatic Change

Lemos MC, Morehouse BJ (2005) The co-production of science and policy in integrated climate assessments. Global Environ Chang 15:57–68. doi:10.1016/j.gloenvcha.2004.09.004

Lemos MC, Kirchhoff CJ, Ramprasad V (2012) Narrowing the climate information usability gap. Nat Clim Chang 2:789–794. doi:10.1038/NCLIMATE1614

Lemos MC, Kirchhoff CJ, Kalafatis SE, Scavia D, Rood RB (2014) Moving climate information off the shelf: boundary chains and the role of RISAs as adaptive organizations. Weather Clim Soc 6(2):273–285

Mauser W, Klepper G, Rice M, Schmalzbauer BS, Hackmann H, Leemans R, Moore H (2013) Transdisciplinary global change research: the co-creation of knowledge for sustainability. Curr Opin Environ Sustain 5(3–4): 420–431

McNie EC (2008) Co-producing useful climate science for policy: Lessons from the RISA Program. Ph.D. dissertation, University of Colorado at Boulder
Melillo, JM, Richmond TC, Yohe GW, eds. (2014) Climate change impacts in the United States: The Third National Climate Assessment. U.S. Global Change Research Program, 841 pp. (NCA 3 report)
Morgan MG, Dowlatabadi H, Henrion M, Keith D, Lempert R, McBride S, Small M, Wilbanks T (2009) Best practice approaches for characterizing, communicating, and incorporating scientific uncertainty in decisionmaking. CCSP 5.2, A report by the Climate Change Science Program and the Subcommittee on Global Change Research. National Oceanic and Atmospheric Administration, Washington
Moser SC, Davidson MA (submitted for this issue) Coastal Assessment: The Making of an Integrated Assessment. Climatic Change
Moss R, Scarlett PL, Kenney MA, Kunreuther H, Lempert R, Manning J, Williams BK, Boyd JW, Cloyd ET, Kaatz L, Patton L (2014): Ch. 26: Decision support: Connecting science, risk perception, and decisions. Climate change impacts in the United States: The Third National Climate Assessment, Melillo JM, Richmond TC, Yohe GW, eds., U.S. Global Change Research Program, 620–647. doi:10.7930/J0H12ZXG
National Climate Assessment. NCA report series, vols. 1–9. All volumes published online by the United States Global Change Research Program, accessible at http://www.globalchange.gov/engage/process-products/NCA3/workshops
National Research Council (2007) Evaluating progress of the U.S. Climate Change Science Program: methods and preliminary results. National Academy Press, Washington, DC
U.S. Global Change Research Program (2011) National Climate Assessment Interim Strategy—Summary (Final) Accessed at http://www.globalchange.gov/sites/globalchange/files/NCADAC-May2011-Interim-Strategy.pdf
U.S. Global Change Research Program (2012) Guidance on information quality assurance to chapter authors of the National Climate Assessment questions tools. Accessed at http://www.globalchange.gov/sites/globalchange/files/Question-Tools-2-21-12.pdf
Waple AM, Champion S, Kunkel K, Tilmes C (submitted for this issue) Innovations in Information Management and Access for Assessments. Climatic Change

Climatic Change (2016) 135:23–37
DOI 10.1007/s10584-015-1501-4

Building a sustained climate assessment process

James L. Buizer[1] · Kirstin Dow[2] · Mary E. Black[1] · Katharine L. Jacobs[1] · Anne Waple[3] · Richard H. Moss[4] · Susanne Moser[5] · Amy Luers[6] · David I. Gustafson[7] · T.C. Richmond[8] · Sharon L. Hays[9] · Christopher B. Field[10]

Received: 10 February 2015 / Accepted: 2 September 2015 / Published online: 21 September 2015

Abstract The leaders and authors of the Third US National Climate Assessment (NCA3) developed new modes of engaging academia, the private sector, government agencies and civil society to support their needs for usable, rigorous, and timely information and better connect science and decision-making. A strategic vision for assessment activities into the future was built during the NCA3 process, including recommendations on how to establish a sustained assessment process that would integrate evolving scientific understanding into decision making to manage the risks of climate change over time. This vision includes a collaborative assessment process that involves partnerships across a diverse and widely distributed set of non-governmental and governmental entities. The new approach to assessments would produce timely, scientifically sound climate information products and processes, rather than

This article is part of a special issue on "The National Climate Assessment: Innovations in Science and Engagement" edited by Katharine Jacobs, Susanne Moser, and James Buizer.

✉ James L. Buizer
buizer@email.arizona.edu

1 University of Arizona, Tucson, AZ, USA

2 University of South Carolina, Columbia, SC, USA

3 Second Nature, Boston, MA, USA

4 Joint Global Change Research Institute, Pacific NW National Laboratory and University of Maryland, College Park, MD, USA

5 Susanne Moser Research & Consulting, Santa Cruz, CA, USA

6 Skoll Global Threats Fund, San Francisco, CA, USA

7 ILSI Research Foundation, Washington, DC, USA

8 Van Ness Feldman, L.L.P, Seattle, WA, USA

9 Computer Sciences Corporation, Falls Church, VA, USA

10 Carnegie Institution for Science, Palo Alto, CA, USA

focusing on the production of single quadrennial synthesis reports. If properly implemented, a sustained assessment would be more efficient and cost-effective, avoiding the painful and time-consuming process of beginning the assessment process anew every 4 years. This ongoing assessment would also encourage scientific and social innovations and explore new insights and opportunities, building the capacity to advance the development and delivery of climate information to meet societal requirements and benefit from scientific opportunities.

1 Introduction

As communities across the nation find themselves coping with evolving climate change impacts, decision-makers are increasingly focused on adapting to both familiar and unexpected challenges. The US federal government has conducted a series of climate assessments to synthesize the state of knowledge about physical climate science and impacts. Historically, the assessments have often been more valuable to the scientific community than decision makers; delivering usable, rigorous and timely information has been difficult. The Third National Climate Assessment (Melillo et al. 2014) developed new modes of engagement between the government, academia, the private sector, and civil society for connecting science and decision-making. The experience of the NCA3 suggests that assessments of both scientific progress and the implications for managing risk, when undertaken on an ongoing and strategic basis, can improve the connections between research agendas and practical applications.

In the initial organizational stages of the NCA3 there was extensive discussion throughout the US climate assessment community about how to improve the overall outcomes of assessment activities. Many in this community had volunteered to participate in past assessments and had reached a broad conclusion that a longer-term, more sustained approach to conducting assessments over time would be more efficient and generate multiple benefits. The vision of building a sustained climate assessment process, now reflected in the Strategic Plan (National Science Technology Council 2012 for the US Global Change Research Program (USGCRP), is aimed at increasing the program's ability to effectively and efficiently support the expanding needs for decision-relevant information.

This vision of a sustained assessment process centers on empowering civil society, the business community, and multiple levels of government with knowledge needed to more effectively manage the risks of climate change. It includes sustained dialogue with users to better understand decision contexts and information needs (and hence novel ways for users to interact with Federal agencies), preparation of a wider range of products, continued innovation in communication of information about climate change risks and opportunities, and additional efforts to build capacity to decentralize assessment across a diverse and widely distributed set of non-governmental entities and multiple levels of government. This assessment process would produce timely, scientifically sound climate information products, systems, and processes to support decision-making across the nation, building the capacity to *advance* the development and delivery of climate information to meet societal requirements and benefit from scientific opportunities.

This article provides background on historical approaches to US National Climate Assessments, lessons learned from them, and the rationale for a sustained assessment process. It concludes with discussion of the challenges involved in building and evaluating a sustained assessment, including suggested metrics of success. The discussion also considers the recommendations of the National Climate Assessment Development and Advisory Committee

(NCADAC) in its Special Report, *Preparing the Nation for Change: Building a Sustained National Climate Assessment* (Buizer et al. 2013). The authors of this article all contributed to that Special Report, and are not disinterested parties.

2 Previous US national climate assessments

The Global Change Research Act of 1990 (GCRA) requires a scientific assessment, including projections of future climate conditions and evaluations of remaining uncertainties, to be completed at least every 4 years by the US Global Change Research Program (USGCRP). However, only three NCA synthesis reports have been completed in the intervening years between 1990 and 2014. There are a number of explanations for why the quadrennial reporting requirements have not been met, including the fact that comprehensive, multi-sector assessments are difficult to conduct; the politics of climate change and funding issues within federal agencies are also factors. Despite earlier efforts to establish an ongoing assessment program, there has been limited institutional support to plan and implement these assessments in an ongoing and strategic manner. The experiences of initiating and conducting the first (NCA1) and second (NCA2) national assessment processes provided valuable lessons that informed the development of NCA3.

2.1 First and second NCAs

The first two NCAs resulted in reports released in 2000–2001 and 2009; they followed different paths in their development and dissemination compared to the NCA3 and to one another. The perspectives of the participants in previous national climate assessments were important in the design of the NCA3 approach, and the lessons learned in the first and second NCA are worth reviewing here.

The first NCA report, "*Climate Change Impacts on the United States: The Potential Consequences of Climate Variability and Change*" (National Assessment Synthesis Team 2000), was a landmark report on climate change for the US, and consisted of a shorter overview (154 pages, published in 2000) and a large 'foundational' document (612 pages, published in 2001). It was the first effort of its kind in the US, and engaged over a thousand contributors and reviewers from academia, government, and public and private sectors. The assessment was developed under the leadership of a National Assessment Synthesis Team (NAST), which was a 14-member committee of experts drawn from academia, government, non-governmental organizations and the private sector. This and all subsequent NCA report development efforts were chartered under the Federal Advisory Committee Act (FACA).

In the development phase of the foundational report (the NCA1 synthesis report), a series of regional and sectoral workshops were conducted, funded by several federal agencies, resulting in reports from 20 regional teams and five sectoral teams and representing an even wider array of contributors. Their reports were reviewed and synthesized by the NAST, and supplemented by additional peer-reviewed material in preparing the foundational document. The synthesis reports were then reviewed by the public and hundreds of experts, including an independent panel formed as a subcommittee of the President's Council on Science and Technology (PCAST). The development and coordination of this effort was undertaken by staff at the National Coordination Office of the USGCRP, under the direction of the Office of Science and Technology Policy (OSTP) in the White House.

The NCA1 synthesis report drew primarily on literature review and input from national and regional assessment meetings. It included a wide array of contributors, and being the first effort of its kind, significant lead-time was required. Process initiation to completion of the overview reports took approximately 6 years; following the publication of the reports in 2000–2001 no further regional or sectoral engagement or outreach strategy was implemented and the infrastructure that was established to provide input to and coordination of the NCA was disbanded. The federal government did not distribute the reports widely. The inability to sustain the assessment effort after the reports were completed has been generally attributed to the transition from the Clinton administration to the Bush administration, which re-organized and reoriented the USGCRP into the Climate Change Science Program (CCSP) and a related program, the Climate Change Technology Program.

Although this first NCA was developed by well-established experts following well-vetted internal guidelines for quality assurance and was extensively and independently reviewed, the Competitive Enterprise Institute sued the government over the NCA1 report in 2003. Their legal challenge was based on the supposition that the NCA1 did not adhere to the recently adopted Data Quality Act (also known as the Information Quality Act). The case was later dismissed.

For the NCA2, a wholly different approach was employed. The strategic plan of the USGCRP (then known as the CCSP; USGCRP 2003) identified five research goals, ranging from extending knowledge of the Earth's past and present climate (emphasizing climate variability), to better characterizing uncertainty, to promoting decision support within the limits of knowledge. A series of 21 topically focused Synthesis and Assessment Products (SAPs), grouped under these strategic goals, were produced between 2006 and 2009 according to guidance criteria provided by the CCSP Coordination Office. These criteria were consistent in some instances with NCA1 (e.g., public review was required), but stakeholder engagement was not required in the development of the SAP reports themselves. It took 6 years to produce all of the individual SAPs, following long discussions about which agencies would be responsible for facilitating which report, how each would be conducted and funded, and when to form FACA committees. The CCSP website documented the progress of each report and made them available for public comments.

The SAPs were scientifically rigorous and wide-ranging in their topical coverage. They were designed to fill important knowledge gaps and answer specific science questions, most of which were priorities for federal agencies and administration policy makers. The SAP reports were written by experts and subjected to relatively transparent review processes, but were not intended to provide a comprehensive national evaluation of impacts. A 2007 lawsuit brought against CCSP seized on this issue and resulted in a judgment that the SAPs did not meet GCRA legal mandates. This led to preparation of a synthesis NCA report (Karl et al. 2009) which was developed within a year by a 31-member committee of federal and academic authors. The process included review by an expert panel and an open review process but limited stakeholder engagement. The Obama administration accepted the report and released the 188-page document, *Global Climate Change Impacts in the United States*, in 2009. The report release was followed by very limited outreach and communication efforts, resulting in the NCA2 not being widely known.

The first two NCA synthesis reports were both of high scientific quality, provided information that filled knowledge gaps, and made information accessible to non-experts. Both assessments emphasized regional findings. The major differences between the two were the degree to which stakeholders were involved in the process and the design of information flow from the underlying reports to the synthesis report. For example, because the NCA2 SAPs did not cover all US regions or even all of the sectors required by the GCRA, the committee placed the SAP findings into a more cohesive regional and sectoral structure. Neither NCA1 nor NCA2 assessment development infrastructure was maintained beyond the release of the synthesis reports.

2.2 The third national climate assessment

From the beginning of the NCA3 process, the USGCRP agencies and the broader science community (through the National Research Council and other venues) were interested in building an efficient, ongoing enterprise of learning and engagement. However, engaging stakeholders, understanding evolving information needs, building on previous regional or topical analysis, and tracking outcomes (and needed improvements) from the previous NCA processes were initially difficult because the staff and support infrastructure were no longer in place. As a result, the NCA3 process was kicked off with a series of "listening sessions" to help shape a strategic plan for the NCA3 and initiate discussions of a sustained assessment process. Public meetings in Chicago in February 2010 included a "Midwest Regional Workshop" and a "Strategic Planning Workshop" (USGCRP 2010a, b).

The Chicago discussions included conversations about the audiences for the assessment, its mission, goals, and principles, and identifying what (limited) assessment activities were already underway within agencies and across the US. Some of the key insights that emerged from the Strategic Planning workshop were: 1) the need for the NCA to support both adaptation and mitigation decisions; 2) the importance of seeing the NCA as both a process and a set of products; 3) the need to do a significantly better job identifying regional issues and engaging with stakeholders than previous assessments had done; 4) the need to support science translation and "boundary-spanning" activities; 5) the importance of establishing metrics that relate to informing policy and supporting decisions; 6) the need to balance "scholarly" contributions with decision relevance; 7) the importance of a phased, sustained approach to ensure timeliness of future products; 8) concerns about finances and how that might constrain the process; 9) the importance of an inclusive vision (reaching outside of the federal agencies); and 10) the need for co-ownership of the assessment activity with external partners. Improving the accessibility of findings and the need to coordinate with the needs and assets of federal agencies were also topics of significant discussion.

Importantly, the Strategic Planning Workshop report (USGCRP 2010b) contained a significant section on "Building an Enduring Assessment Structure," including the following priorities (paraphrased from the workshop report):

- Identify and engage stakeholders from the beginning
- Identify and prioritize stakeholders' needs
- Map existing capacity and capabilities
- Move from a focus on vulnerability assessment to building resilience
- Develop data, observations, and trusted sources
- Develop a strategy for communication early in the process

The discussions at this early workshop, the development of the actual NCA3 process, and the final recommendations of the federal advisory committee (NCADAC) in the Special

Report on how to sustain the assessment process (Buizer et al. 2013) are remarkably consistent, with a strong focus on the value of building a sustained assessment process.

3 Benefits of a sustained assessment process

The changes underway in climate and weather systems are driven by a range of complex processes and feedbacks. Managing the increasing pace of change and the associated consequences requires actionable, up-to-date information. Although scientific understanding of the climate system is improving, resource managers, businesses and private citizens often do not have the information they need to support decisions about managing greenhouse gas emissions and managing the risks associated with changes that are already occurring, let alone those projected for the future. There is demand for data at the scale of decisions – often this means at the community scale – and for more timely access to credible and actionable scientific information.

The primary benefit of a sustained assessment process is the establishment and maintenance of ongoing dialog among potential users and producers of scientific information needed to support national and global efforts to manage risks to both people and planetary life support systems. An ongoing, distributed assessment process can leverage improvements in both scientific rigor and the salience of findings. It can provide pathways for moving the science produced under the auspices of the USGCRP to civil society, the private sector, state/local governments, and regional entities who can apply this information to evaluate the potential implications of global change for their own interests. Without the establishment of a sustained process that harvests information from multiple knowledge systems and channels, information would need to flow through a centralized, government-driven process, which is less likely to be able to meet these diverse information needs.

4 Building a sustained assessment process

From its formation, the NCADAC was tasked with both developing the NCA3 synthesis report and providing advice to the government on how to build a sustained assessment process. According to its charter, "The committee's mission is to synthesize and summarize the science and information pertaining to current and future impacts of climate change upon the United States; *and to provide advice and recommendations toward the development of an ongoing, sustainable national assessment of global change impacts and adaptation and mitigation strategies for the Nation*" (emphasis added). Accordingly, the NCADAC decided to include a discussion of the rationale for a sustained assessment process as a chapter of the final NCA3 synthesis report and also to prepare a Special Reportadvising the government on how and why such a process should be established.

4.1 A new vision and process for climate change assessments: special report on the sustained assessment

Preparation of the special report was assigned to an interdisciplinary subcommittee of NCADAC members. The authors drew on a number of National Research Council reports, reviews of prior assessments, and their own experiences interacting with users, Federal officials, and members of the research community.

The recommended vision for the sustained assessment is creating.

> "…an **inclusive, broad-based, and sustained process** for assessing and communicating scientific knowledge of the vulnerabilities, impacts, risks, *and opportunities* associated with a changing global climate in support of decision-making across the United States.

supporting the goal to.

> "Enhance the ability of *decision-makers at multiple scales throughout* the United States to **anticipate, mitigate, and adapt** to changes in the global environment."

Clearly this vision required something more than continuous production of NCA- and IPCC-like reports. In fact, at its core, the goal and vision for the sustained assessment involved broadening the assessment process beyond the Federal government – to engage and empower decision makers throughout society to use the best available scientific information to come to terms with climate-change risks, opportunities, and uncertainties. And, as described in the report, it envisioned making a more diverse set of products available (including data sets, maps, indicators, evaluations of decision support science, and others) through various communication mechanisms involving ongoing interactions among producers and users of scientific information.

The Special Report (Buizer et al. 2013) was formally submitted to the government in October 2013, in advance of issuance of the final synthesis report. It suggests how to develop a more efficient and strategic ongoing process, as well as programmatic approaches and investments that could significantly enhance both the utility and the scientific rigor of future processes. As the Special Report notes,

> "a sustained process offers the opportunity for planning and investment decisions to be more deliberate and phased in over time…(allowing)…the US government (to) more efficiently support the science and adaptation needs of federal agencies; and provide transparent access to data at a variety of scales for private businesses, local/state/regional/tribal governments, and other organizations that are planning for the future." (p.7)

It also points out that "a strong federal commitment to documenting and anticipating both the positive and the negative aspects of climate and global change demonstrates leadership that can further encourage broad non-governmental engagement….and allow greater efficiency in development of assessment products" (ibid.).

The specific recommendations of the Special Report, grouped into four categories below, are discussed in detail in the report, along with criteria for prioritization of assessment efforts.

a) *Establish mechanisms to support enduring collaborative partnerships that sustain assessment activities*: this was seen as a central challenge because it requires USGCRP and participating agencies to develop unprecedented long-term relationships between the research community and decision makers. The report describes engagement, communication, and partnership opportunities that provide 'co-production' capacity.
b) *Enhance and organize the scientific foundations for managing the risks and opportunities of climate change*: this section focuses on integrating fundamental scientific knowledge with decision support processes to develop new products and tools that support

assessments and create new knowledge systems to link scientists and information users. These products and systems include:

- methods for vulnerability assessment and risk management;
- development of indicators of change, scenario methods and products, and valuation methods;
- ways to incorporate climate-related international influences on the US;
- methods for assessing confidence and uncertainty;
- adaptive learning within assessment processes;
- identification of risk management information needs

c) *Provide infrastructure to support a sustained assessment process*: this section describes a variety of 'infrastructure' needed for sustained assessment including good leadership and a strong coordination office, processes to support preparation of several different types of reports, data and information management systems, and regional institutions and networks; and
d) *Diversify the resource base and set priorities*: this was seen as both an opportunity and necessity – establishing an opportunity to draw on a wide range of resources from the private sector and civil society; it is also viewed as a necessity in light of constrained federal resources.

4.2 Innovations of NCA3 that inform a sustained assessment process

The NCA3 process included a number of innovations that provide valuable ideas and insights for transitioning to a sustained assessment. These included development and promotion of guidelines to encourage 'risk-based framing' to identify climate-related impacts of high consequence; attempts to improve assessment and communication of levels of confidence and improve transparency of author team deliberations through preparation of 'traceable accounts' that explain the author's thought process and sources for key findings; and development of scenario approaches that incorporated lower probability events (for sea-level rise) and that encouraged participatory scenario planning to explore implications of uncertainty. Several of these innovations are discussed in other chapters of this volume. We emphasize here two innovations related to expanding the information base for the assessment and improving online delivery of information and access to underlying data:

4.2.1 Peer-reviewed publications vs. information quality act innovations

In the effort to enhance the relevance of process and products for decision-makers, the NCA3 authors were asked to go beyond the standard academic literature where necessary or possible to illustrate impacts, provide case studies, or integrate important new insights. In many instances they found government documents and other sources of highly reviewed information that were viewed as credible by subject experts; these were relied upon to support important conclusions. In other cases, the authors found information that was extremely useful in case studies and other illustrations of climate change impacts and responses, but considered less reliable from the perspective of supporting scientific conclusions themselves.

In anticipation of concerns about the use of non-traditional sources of information for the NCA3, a separate subcommittee of the NCADAC developed guidance on how to manage the requirements of the Information Quality Act while incorporating some non-traditional literature (USGCRP 2011). This guidance was crucial and helped set expectations about sources of information for all the authors and the government reviewers. There were occasions during the review process when specific sources were challenged, but in all cases a resolution was found that allowed the findings to be included appropriately. This is particularly important in light of proposed long-term partnerships with non-governmental external parties who would like to play a role in the ongoing process, and who particularly value some sources of information (such as information about the current status and success of adaptation projects) that is unlikely to be updated on a regular basis through the peer-reviewed literature.

4.2.2 Online delivery and transparent access to underlying data

A dramatic change between the NCA3 and its predecessors was its electronic delivery via an interactive (and attractive) website. This meant that all of its contents were searchable online and that all of the evidence behind the findings could be linked rather than just cited in a bibliography. The fact that the interagency Global Change Information System was built by USGCRP as a means to support and deliver the NCA3 synthesis report marks a major transition to a new era of information access that reaches far beyond the NCA because it can facilitate ongoing data-sharing and analysis across agencies and support subsequent NCA reports. Advances in information technology, information systems, author support platforms, and web-based search functions used in the NCA3 process have permanently changed the way assessments will be conducted. Sustained assessment is much more viable in the context of automated submissions, online review, electronic reports, and high-volume data-management systems (Waple, Submitted for publication in this special issue).

5 Such a good idea – so hard to implement!

There are many barriers to creating a permanent, sustained assessment process; the majority of them relate to concerns about the word "sustained." Under current federal budget constraints, it is hard to agree across multiple federal agencies to any kind of ongoing expenditures, even for a program that is congressionally mandated. Some federal agencies and program managers within the USGCRP have expressed concern about exactly what a sustained assessment might entail. The word "sustained" could be understood to mean maintaining the same level of effort required for the extremely involved NCA3 process and its very large participant list. However, the Special Report (Buizer et al. 2013) and the USGCRP Strategic Plan (National Science Technology Council 2012) note that efficiencies can be achieved through a well-planned, ongoing process, while also improving products, regardless of the size of specific assessment efforts. There are also uncertainties about the extent of the ongoing role of non-federal participants and contributors in a process that is fundamentally a government responsibility, especially given the constraints of FACA, which governs the degree to which non-federal groups can provide consensus advice to the government on an ongoing basis. Other barriers to progress include issues associated with leadership, resistance to change, and governance issues.

5.1 Budgetary constraints

There has never been a budget line for the USGCRP Coordination Office itself or the NCA process, even though preparation of these reports is a federal mandate for the USGCRP. At the start of the NCA3 effort (in FY 2010) funding was identified by the Office of Management and Budget within one agency (NOAA) as its contribution to the collective assessment process, but other agency contributions were not specified. In part, the need for building financial support across the agencies has a positive effect – it reinforces their "ownership" of the processes and products jointly created. However, having no explicit interagency budget line for the NCA means that existing agency programs need to be leveraged and/or "taxed" to support the assessment. The lack of sufficient ongoing funds to support the sustained assessment remains a significant challenge.

Importantly, federal program managers operate in an environment of constantly increasing expectations on a fixed (or in some cases, decreasing) budget. In this context, it would be understandable if they saw investments in the NCA as one more unfunded mandate. It is much easier to start new programs that are additive (bringing in new resources) than to engage in a zero sum game. Understandably there was some reluctance to fund NCA3 activities under highly constrained fiscal conditions. Despite this challenge, federal managers involved in the NCA3 clearly embraced the general role of assessments in the scientific process.

Based on personal communication with program managers, there are clear differences between internal assessments of agency program outcomes and broad-scale assessments that are highly vetted like the NCA3. The multiple levels of review in NCA processes add significant credibility to the outcomes. In the NCA3, the number of topics involved, the array of participants, the intersections of physical and social science, the multiple geographic scales of evaluation, and the time frames for future projections mandated by law all added cost and complexity. But the benefits of the NCA3 process were well recognized by federal leadership, particularly by those who represented their agencies in the interagency NCA working group that helped build and manage the process on a day-to-day basis.

5.2 Losing control

One barrier to conducting highly transparent, broad-scale assessments with significant stakeholder engagement is the potential for loss of control for the federal participants. Including on-the-ground managers and stakeholders from regions and sectors in assessment processes leads to new sources of information that may challenge conventional approaches to science and to its interpretation. For example, some federal agency representatives expressed concerns during the development of the NCA3 process that, given the political nature of climate issues, strong engagement with stakeholders could lead to a potential loss of control over the process itself. Many federal program managers have had negative experiences in public meetings with confrontational individuals, inaccurate press reports, or other consequences of poorly designed engagement strategies or unforeseen events causing unexpected outcomes. There is justifiable anxiety about government and scientific processes that are conducted in a truly public arena. It is not surprising that there could be reluctance to engage in a major way with stakeholders.

However, there is also broad acknowledgement that the federal government's conduct of comprehensive assessments can and does benefit from the input of external parties and on-the-ground knowledge. Multiple previous reports (e.g., NRC National Research Council 2007, National Research Council 2009, National Research Council 2010) have noted that assessments benefit from more interactive, inclusive processes. The federal government simply does not have all of the kinds of expertise required to evaluate the risks and opportunities associated with climate impacts. However, broadening the assessment effort to incorporate multiple sources of knowledge, including traditional ecological knowledge and the perspectives of private and public sector managers within regions and sectors, is challenging (see Jacobs and Buizer, this issue).

Possibly the most important scientific opportunity associated with expanded participation in assessment activity is the potential change in scientific understanding that can come from sharing information across multiple scientific disciplines and practitioner perspectives, access to new data sources, and changed research agendas. The trend toward integration of multiple kinds of knowledge has led to activities such as "Integrated Assessment Modeling" that work towards predictions of future conditions while taking into account many different sources of data and knowledge. Risk-based framing, an interdisciplinary approach that considers biological, social, physical, and health impacts, helped NCA3 authors identify gaps in knowledge that need attention from the scientific community. While such approaches can challenge the views and investments of more conventional science, the resulting improvements in scientific understanding have great potential for societal benefit.

5.3 Partnership opportunities and complexities

As recommended in the Special Report, any ongoing assessment process will need to diversify its resource base. A shrinking domestic federal budget, along with expanding demands for services, implies that changes are required.

An ever-increasing number of foundations, private companies, and NGOs are working on climate issues and investing funds in research, education, and communication. Communities that are actively engaged in managing risks are interested in working with the federal government to ensure that assessment processes provide the kinds of data they find most useful and are willing to provide in-kind services or even financial assistance. The amount of activity focused on adaptation planning (Bierbaum et al. 2014) is also increasing. Although the NCA3 found that the level of adaptation activities occurring is not commensurate with the need or future challenges from climate change, it is clear that the interest level in more and higher quality climate information is rising. Using only the measure of the number of hits on the NCA3 website in the first two months since its release (1.5 million) by comparison to previous hits on the USGCRP website following previous report releases (orders of magnitude lower), gives an indication of the expansion of interest.

Because of resource constraints there is a need to leverage existing investments and seek opportunistic approaches that lead to win-win solutions. For example, future quadrennial NCA synthesis reports may not actually drive research agendas, but they certainly can harvest information from agency documents as it becomes available. A properly designed sustained assessment process encourages agencies and contributors to conduct activities and produce products to meet their own needs that are also useful in a subsequent synthesis report.

The Special Report (Buizer et al. 2013) recommends expanding the partnerships that were initiated in the NCA3 process, also noting that doing so might require shared governance of some aspects of the process. The federal government has reason to be cautious about this approach, given the important political and regulatory implications of climate products. In addition, the scientific community has reason to be concerned about the possibility of interference with the process or impacts on the credibility of the findings from some future partnerships. It is imperative that the credibility of the assessment process be untarnished, yet non-federal resources are likely to be a critical component of most paths to a sustained and useful assessment. A well-designed system of shared governance and information quality assurance can support the construction of credible outcomes from distributed processes that involve both federal and non-federal resources and input, but it will require careful attention to both the appearance and the reality of avoiding potential conflicts of interest.

5.4 Challenges requiring resolution in establishing a sustained assessment

There is natural tension between the need for government support and engagement in assessment processes and the need for independence from the government, politics, FACA rules, etc. Government often moves slowly and cautiously, while the private sector and non-governmental entities can often shift direction and priorities more quickly. If true partnerships are to emerge, short-term political and economic considerations need to be less prominent than the longer-term needs of the country, the scientific community, and civil society. At some level, the federal government must support a sustained assessment process, not only because of GCRA requirements but because of its own need for accurate and integrated scientific information to support research and decision-making. Clearly the sustained assessment process must meet the needs of the federal government, its major stakeholder, in order to succeed.

However, over more than two decades since the GCRA was passed, the federal government has demonstrated how difficult it is to move past precedent and historic ways of conducting assessments. Ending tiresome debates about process issues and focusing instead on "the art of the possible" is important. For example, the government may need to be both more strategic and more opportunistic in order to leverage work initiated for other purposes that is timely, credible, and useful to stakeholders and the sustained assessment process. It will never be possible to make all assessment process decisions in advance, but rather, a key to success is learning from ongoing efforts and collaborating to identify ways contributions will be most useful. In addition, the government should provide basic guidance to the scientific and user community that facilitates participation – e.g., by providing guidance on how an external party can contribute data or reports for consideration in a future assessment, including how to document data sources, processes, and conclusions so that the products can be more easily used.

6 Measuring success of a sustained assessment process

One way to test whether a sustained assessment is successful is to evaluate progress over time based on established criteria (the Special Report includes suggestions for such criteria). We offer additional criteria here for consideration. For example, can future report processes meet the four-year deadline? This is one (admittedly limited) way to test the Special Report's assertion that a standing advisory committee, an ongoing set of interim reports to harvest from, a well-trained staff, and well-defined external partnerships can result in a more efficient process. Second, is there a documented increase in demand for products, including the full range of electronic and hard copy products as well as the less traditional products and data that are expected to be of interest to decision-makers? Third, is the engagement enterprise widely perceived as successful? This would include a review of the expectations and performance of external partners and funders in a broadening array of partnerships, as well as internal engagement efforts. For example, a metric could be the number and quality of partnerships, measured in terms of numbers of people engaged, types of engagement, and documented applications of the information produced. And fourth, is the process linked on an ongoing basis to federal and external partners' research agendas?

A critical component of success in the sustained assessment process is whether it is perceived to be truly owned and supported by its parent organization, the USGCRP, relevant USGCRP agencies, and the named partners in the process. Evidence could include ongoing support for developing indicators of change, development of useful and timely scenario products, or improvements in the functionality of the website and the data management system. Metrics could also include staffing levels, agency use of assessment products, and the quantity and quality of technical input documents submitted for consideration. To assess success in the eyes of external parties, evaluation by independent NCA partners and information users should become an integral component of the process.

Different measures of success may be needed over different time frames, from the short term to the long term. Metrics could focus on issues related to process, outputs/products, and outcomes. Agreeing on what success looks like in a broad, multi-party, multi-objective process is very difficult. Key considerations include:

- One measure of success would be to examine whether the NCA process is producing a more diverse set of products – data sets, maps, targeted information on extreme events, evaluations of decision support systems and processes, user forums, and others. Reports will, of course, continue to be a mainstay of the process, but as discussed above (and in Moss, this issue) the growing range of decision contexts and information needs requires an expanded product set.
- After 10 years, success could mean that multiple people and organizations across regions and sectors have used NCA products in their own assessments and data to make decisions. If the sustained assessment is successful, assessment and decision making processes associated with the NCA will be more widely distributed, and tracking this evolution will require ongoing scholarly work demonstrating that the sustained assessment has played a significant role in how decision-relevant science is developed and used in decisions. These metrics of success would require a process to be in place that allows monitoring and study of the process of conducting and building the sustained assessment.
- Measures should capture whether the ongoing process is inclusive and has a more diverse set of players over time. Metrics could include whether states or other countries are following the NCA model; whether local governments, industry and philanthropy are engaging in funding, knowledge creation, and data-sharing; whether ongoing private sector relationships are built around the NCA; and whether a self-identified and self-organized community continues to engage and be part of the assessment process.
- Measures of the degree to which USGCRP and the federal member agencies have embraced the process as a central component of its program activities could be established across agencies, including whether components of the sustained assessment process continue to show up in the strategic plans and budgets of USGCRP; and whether academic partners, private sector interests, agencies and program managers are able to get funding for projects that support "sustained assessment" activities.
- Success can also be measured in terms of engagement of the assessment community, its collective capacity, and its sense of shared accomplishment, whether the premier scientists and stakeholders continue to choose to spend their time on assessment, and whether the

science gaps noted in the Special Report and subsequent assessment activities are starting to be filled.

Ultimately, the success of a sustained assessment should be measured relative to outcomes. If a goal is a more resilient society, an important measure would be linking information in the NCA reports to use of information and, ultimately, to evidence of reduction of risks. In the short term, measures could be more rudimentary; for example, are the agencies and external partners actually using NCA information in managing risk? The US government's ongoing public commitment to an ongoing assessment process would also qualify as a success.

7 Conclusions

The experience from the past three national climate assessments suggests that moving toward a sustained assessment model would provide efficiency and effectiveness in responding to decision makers' information needs for climate risk management. It would also enable capacity building that supports climate adaptation and mitigation, encourages innovation, provides new interdisciplinary scientific insights and opportunities, and ensures greater utility of future NCA findings. Establishing an adaptive process designed to test new approaches and continuously evaluate them would improve both the scientific content and the utility of the information products. The NCA3 experience demonstrates that an inclusive approach to assessments can lead to more real-time participation and decision-relevance. Increased engagement, better representation of sectors and scales, political, social and geographic diversity, and a more integrated community of scientists and practitioners who can work together to solve issues of concern to society can all contribute to better risk management strategies.

The NCADAC Special Report recommends the following steps for building an effective sustained assessment process: 1) build mechanisms to support collaborative partnerships, 2) develop the scientific foundations for improving assessments over time, 3) provide adequate and enduring infrastructure (including leadership and staffing), and 4) develop a diversified resource base within and beyond the federal government. There are many successes from previous assessments, but making consistent progress will require trying new approaches, as recommended in these four steps. A properly designed sustained assessment process would *advance* the development and delivery of information in ways that society demands, to manage the risks of the changing climate.

USGCRP faces decisions about the structure, leadership, and scientific underpinnings of ongoing assessments in order to ensure credible outcomes that are useful for managing risk while also meeting the needs of the federal science agencies and broader user community. It is not yet clear whether the USGCRP will seize the current opportunity and use the momentum of the NCA3's success and lessons learned to ensure that its research investments continue to meet the needs of people nationally and internationally, or whether a wide range of factors will be allowed to limit progress. The lessons documented in this special issue provide a foundation for future climate assessment in the US and elsewhere.

Acknowledgments The authors are grateful for important suggestions made by the reviewers, who suggested a number of important improvements.

References

Buizer JL, Fleming P, Hays SL, Dow K, Field CB, Gustafson D, Luers A, Moss RH (2013) Report on preparing the nation for change: Building a sustained national climate assessment process. Committee, National Climate Assessment and Development Advisory

Bierbaum R, Lee A, Smith J, Blair M, Carter LM, Chapin III FS, Fleming P, Ruffo S, McNeeley S, Stults M, Verduzco L, Seyller E (2014) Ch. 28: Adaptation. Climate change impacts in the United States: The third national climate assessment, Melillo JM, Richmond TC, and Yohe GW, eds., U.S. Global Change Research Program, pp. 670–706. doi:10.7930/J07H1GGT

Global Change Research Act of 1990. Public Law 101-606 [S. 169]; November 16, 1990, 104 Stat. 3096-3104

Karl, TR, Melillo JM, Peterson TC, eds. (2009) Global climate change impacts in the United States: 2009 Report, Cambridge University Press. 188 p. http://nca2009.globalchange.gov/download-report

Melillo JM, Richmond TC, Yohe GW, eds. (2014) Climate change consequences in the United States: The third national climate assessment, U.S. Global Change Research Program, 841 pp.

National Assessment Synthesis Team (2000) Climate change impacts on the United States: The potential consequences of climate variability and change, report for the US global change research program. Cambridge University Press, Cambridge UK, 154 pp.

National Assessment Synthesis Team (2001) Climate change impacts on the United States: The potential consequences of climate variability and change, report for the US global change research program. Cambridge University Press, Cambridge UK, 620 pp.

National Research Council (2007) Board on atmospheric sciences and climate. In: Analysis of global change assessments: Lessons learned. National Academies Press, Washington

National Research Council (2009) NRC Committee on Strategic Advice on the U.S. climate change science program. Restructuring federal climate research to meet the challenges of climate change. National Academies Press, Washington

National Research Council Council NR (2010) NRC panel on informing effective decisions and actions related to climate change, informing an effective response to climate change. National Academies Press, Washington

National Science Technology Council (2012) National global change research plan 2012–2021: A strategic plan for the U. S. Global Change Research Program http://downloads.globalchange.gov/strategic-plan/2012/usgcrp-strategic-plan-2012.pdf

USGCRP (2003) Strategic plan for the US Global Climate Change Science Program, at http://www.globalchange.gov/about/strategic-plan-2003/2003-strategic-plan

USGCRP (2010a) In: The United States National Climate Assessment NCA report series, vol 1: Midwest Regional Workshop: Feb. 22-24, 2010, Chicago. Illinois, USGCRP

USGCRP (2010b) The United States National Climate Assessment NCA report series, vol. 2: Strategic Planning Workshop. USGCRP

USGCRP (2011) Information Quality Principles, at http://www.globalchange.gov/sites/globalchange/files/Information-Quality-Principles-Draft_2011-11-16.pdf

Waple, A. (submitted for this issue) Innovations in Information Management and Access for Assessments. Climatic Change

References

Climatic Change (2016) 135:39–54
DOI 10.1007/s10584-015-1568-y

Engagement in the Third U.S. National Climate Assessment: commitment, capacity, and communication for impact

Emily Cloyd[1,2] · Susanne C. Moser[3,4] · Edward Maibach[5] · Julie Maldonado[6] · Tinqiao Chen[7]

Received: 9 June 2015 / Accepted: 19 November 2015 / Published online: 25 November 2015

Abstract The National Climate Assessment's ability to support decision-making partly relies on engaging stakeholders throughout the assessment process. The guiding vision for the Third National Climate Assessment (NCA3) was for an inclusive, broad-based, and sustained process attentive to both the conduct of assessments and communication of findings. Such a process promotes dialogue between scientific experts, stakeholders, and decision-makers about what is important in a particular region or sector, the potential impacts of climate change, and possible responses. We sought to create actionable research and assessment products widely perceived as credible, salient, and legitimate. The process also sought to build capacity to conduct sustained assessments and use climate change information in decision-making processes. Here we describe how we pursued this stakeholder engagement vision during the planning, development, and release of NCA3. Through repeated opportunities for stakeholder. input, we ensured process transparency and inclusiveness in the framing of assessment and built human capital. We also increased connectivity among stakeholder organizations. By

This article is part of a special issue on "The National Climate Assessment: Innovations in Science and Engagement" edited by Katharine Jacobs, Susanne Moser, and James Buizer.

Electronic supplementary material The online version of this article (doi:10.1007/s10584-015-1568-y) contains supplementary material, which is available to authorized users.

✉ Emily Cloyd
ecloyd@usgcrp.gov

1 Present address: ICF International, Washington, DC, USA

2 University Corporation for Atmospheric Research, Washington, DC, USA

3 Susanne Moser Research & Consulting, Santa Cruz, CA, USA

4 Woods Institute for the Environment, Stanford University, Palo Alto, CA, USA

5 Center for Climate Change Communication, George Mason University, Fairfax, VA, USA

6 University of California, Santa Barbara, CA, USA

7 Michigan State University, East Lansing, MI, USA

cultivating a network of collaborators who connected the NCA to other networks, the NCA3 engagement process laid the groundwork for a sustained assessment - which is envisaged to transition the traditional quadrennial assessment approach into a more dynamic and adaptive assessment process.

1 Introduction: engagement and communication as central elements of assessments

Over the past few decades, there has been significant progress in understanding the physical climate system and in documenting impacts of climate change on social-ecological systems (National Research Council [NRC] 2010a; NRC 2007a). Scientific assessments, such as the National Climate Assessments conducted by the U.S. Global Change Research Program (USGCRP) (e.g., Karl et al. 2009; Melillo et al. 2014a) and the Intergovernmental Panel on Climate Change (e.g., IPCC 2014), seek to integrate such scientific information to better inform decision-making (Keller 2010; NRC 2007a; Farrell et al. 2006, Jäger and Farrell 2006). Yet assessments by themselves have not necessarily resulted in greater awareness of climate change risks among citizens and policy-makers or in decisions that explicitly incorporate climate change; to be useful in decision-making, assessments must be accessible and responsive to the needs of users (Moss 2015; Moss et al. 2014; Dilling and Lemos 2011; NRC 2010b, 2008, 2007a and the vast body of literature cited therein).

The Global Change Research Act of 1990 (PL 101-606, Section 106) requires that national climate assessments be produced, but does not specify any requirements for stakeholder engagement. It does charge the USGCRP to "consult with actual and potential users of the results of the Program to ensure that such results are useful in developing national and international policy responses to global change" (Section 102(e)). In addition, the Data Quality Act of 2001 and subsequent Office of Management and Budget guidance about the review process for "highly influential scientific assessments" includes public participation as a component of "process integrity" (70 FR 2664).

The Third National Climate Assessment (NCA3) made stakeholder engagement a principal pillar of the assessment process and sought innovative ways to make the assessment more accessible. This decision resulted from the lessons of previous assessments. The First National Climate Assessment (NCA1, conducted from 1997 to 2000) included strong regional and sectoral stakeholder engagement from the start. This mostly took the form of a series of regional workshops in which stakeholders identified priority concerns, contributed specialized expertise, and identified potential response options (USGCRP 2015). The resulting regional chapters and full report reflected stakeholder concerns to some extent. Members of the Federal Advisory Committee convened to produce NCA1 and outside evaluators of the process recognized the essential role engagement played in creating an effective assessment and noted that continued engagement of a wide variety of scientists, managers, decision-makers, and other stakeholders would be vital to continued success (Parson et al. 2003; Morgan et al. 2005; Moser 2005; NRC 2008). However, there was no Federal support for such ongoing stakeholder involvement, or for outreach and engagement following the release of NCA1. For a limited period of time, outreach was undertaken by a coalition of non-governmental organizations (Moser 2005).

The second assessment did not sustain the level of engagement seen in NCA1; instead, it was primarily a synthesis of 21 scientific reports (called Synthesis and Assessment Products) published by the Climate Change Science Program (CCSP) from 2006 to 2009. There were some opportunities for participation in the development of these documents: the selection of topics was motivated in part by public input (CCSP 2003), and all reports, including the 2009 synthesis that became the Second National Climate Assessment (NCA2), went through a public comment process. However, the level of stakeholder engagement was left to the discretion of individual agencies that led each report (NRC 2007b). At the end of NCA2, the authors advanced a vision for "sustained, extensive stakeholder involvement" in future assessments, noting that the "value of stakeholder involvement includes helping scientists understand what information society wants and needs…the problem solving abilities of stakeholders will be essential to designing, initiating, and evaluating mitigation and adaptation strategies" (Karl et al. 2009, p. 158).

For NCA3, USGCRP convened a National Climate Assessment and Development Advisory Committee (NCADAC), under the sponsorship and auspices of the National Oceanic and Atmospheric Administration (NOAA), to produce the report and to provide advice on the sustained assessment process (Jacobs and Buizer 2015).[1] In spring 2011, the NCADAC approved an interim strategy for the assessment process that included an overarching goal "to enhance the ability of the United States to anticipate, mitigate and adapt to changes in the global environment" (NCADAC 2011a, p. 1) and a vision for "an inclusive, broad-based, and sustained process for assessing and communicating scientific knowledge of the impacts, risks and vulnerabilities associated with a changing global climate in support of decision-making across the United States" (NCADAC 2011a, p. 2). The strategy stated that "an engagement strategy that leverages science and assessment capacity across the United States, while ensuring that the NCA process and products are accessible and useful to stakeholders and the general public, is critical to this vision" (NCADAC 2011a, p. 2); the engagement strategy detailing the approach was also approved (NCADAC 2011b).

One important way that this commitment to engagement and communication became manifest is the inclusion of two communication experts on the NCADAC who co-led a working group on engagement and communication.[2] Engagement efforts were discussed at almost every NCADAC meeting and carried into the ongoing proceedings of the Executive Secretariat of the NCADAC. In addition, USGCRP had one full-time staff member dedicated to engagement, communication, and partnership building who was responsible for ensuring implementation of the strategic advice from the NCADAC.

Below we describe guiding principles of stakeholder engagement for the NCA3, and the communication, engagement, and network-building that occurred during its development, at the report release, and afterward. We discuss which communities were engaged and the collaborations that were formed to help with communication and engagement and conclude with a brief assessment of impact and larger lessons for the sustained assessment.

[1] The NCADAC charter is available at http://downloads.globalchange.gov/nca/NCADAC/NCADAC_Charter_6-24-13.pdf.

[2] This working group was also charged to focus on how to design and embed ongoing evaluation of the entire NCA process into the assessment process, until the topic was deemed to deserve its own working group. Evaluation was eventually included as a recommendation in the sustained assessment special report (Buizer et al. 2013).

2 Principles guiding effective engagement

For NCA3, engagement was defined as an organized process that provides individuals and organizations with access to the design, assembly, content, and products of the NCA by means of two primary and related vehicles (NCADAC 2011b):

- *Communication*: Methods of alerting and informing individuals and organizations about the NCA process and products with the aims of increasing people's interest in and understanding of the NCA, climate change, and the implications of a changing climate for the US, increasing participation, and encouraging use of assessment findings;
- *Participation*: Methods of providing individuals and organizations with opportunities to actively contribute to the assessment – through written inputs and participation in assessment activities – with the aims of increasing the assessment's quality, responsiveness, and utility.

The development of the NCA3 engagement strategy was guided by four overarching principles derived from extant literature (e.g., Dilling and Lemos 2011; NRC 2008, 2007a; McNie 2007; Jäger and Farrell 2006) and experience of the NCA3 leadership:

- *Early and often*: Multiple and varied opportunities were offered for participation during all stages of the assessment process;
- *Inclusive*: Contributions were sought from a diverse set of stakeholders; groups beyond those who contributed to assessments in the past (i.e., scientists, Federal agency experts, and non-governmental groups interested in climate change) were proactively engaged;
- *Sustained*: Relationships with stakeholders – and networks of stakeholders – were developed and maintained indefinitely beyond the NCA3 release; and
- *Enabling*: Capacity was built in myriad organizations beyond the Federal government for individuals to contribute to, use, and communicate assessment findings.

The breadth and number of potential stakeholders for a national climate assessment is considerable. To organize the approach to stakeholder engagement, collaborators were prioritized by type (e.g., government, private sector) and scale of action (e.g., local, regional) (Supplementary Material 1).

3 Planning for engagement: process, strategy and internal organization

The NCA3 engagement effort required considerable internal resources and planning. This planning was undertaken by NCA staff and the NCADAC Engagement Working Group and encompassed both Federal and non-Federal components (Figs. 1, 2). Key elements are described below.

Dedicated staff From the beginning of her tenure in January 2010, the NCA Director made stakeholder engagement a priority. Early workshops on regional impacts of climate change and strategic planning (held in February 2010) included participants representing both contributors to previous NCAs and people and organizations who were new to national assessments. By

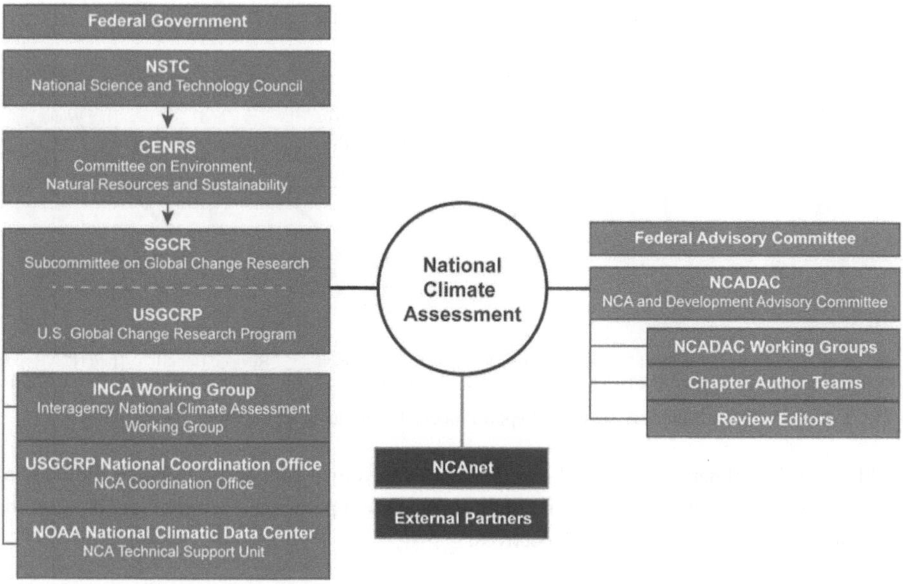

Fig. 1 Organization of NCA components. From Melillo et al. 2014b

summer 2010, the NCA staff included a full-time public participation and engagement coordinator to lead development and implement the NCA3 engagement plan.

Building diverse teams within government In March 2010, USGCRP formed a new interagency working group, the Interagency National Climate Assessment (INCA) Working Group. INCA coordinated, supported, and implemented the Federal components of the NCA, including an interagency operational plan for the NCA, development of technical products, and leadership of expert and stakeholder workshops.[3] Building on the call for broad engagement, INCA members made specific efforts to recruit additional participants from agencies outside of the core USGCRP participants. In some cases, that meant program managers whose cabinet-level departments participated in USGCRP but whose agencies were less engaged (e.g., within the Department of the Interior, the National Park Service, Bureau of Indian Affairs, Bureau of Reclamation, and the U.S. Fish and Wildlife Service) and in other cases, agencies with clear interests in climate change that had not previously been part of USGCRP (e.g., the Department of Homeland Security's Federal Emergency Management Agency).

Explicit focus on engagement within the NCADAC In late 2010, the Department of Commerce announced the formation of the NCADAC. The solicitation of nominations for this advisory committee noted the need for "a very wide range of expertise" and called for individuals "with experience in private industry, state, local, and regional government, academia, and non-governmental organizations, [...] drawn from a broad geographic distribution" (76 FR 11427). To accommodate this diversity, the NCADAC consisted of 44 non-Federal members and 16 Federal ex officio members. Among the various working groups formed under the NCADAC, one focused on engagement and included both NCADAC members and

[3] More information about INCA is available from http://www.globalchange.gov/about/iwgs#INCA.

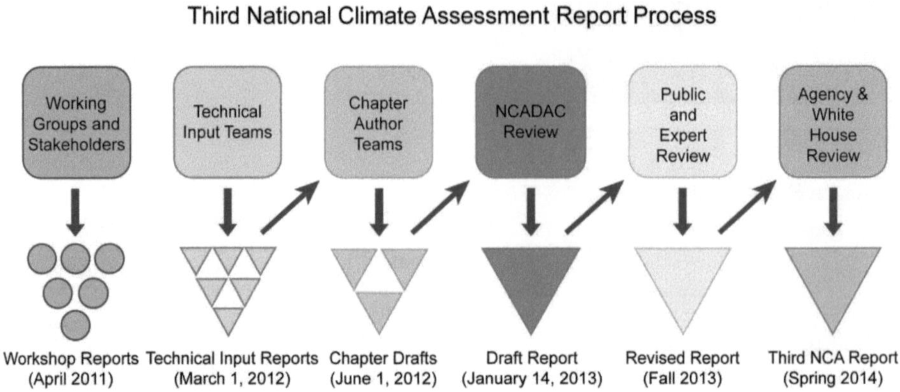

Fig. 2 Third National Climate Assessment Report Process. From Melillo et al. 2014b

additional disciplinary experts and assessment users from within and outside of the Federal government. This helped ensure that communication and participation were carefully considered and informed by multiple perspectives.

Early lessons in engagement When the USGCRP initiated the NCA3 process, a primary guiding principle for the assessment was to "maximize engagement of stakeholders" (USGCRP 2010a). This was reflected in process and methodology workshops in 2010 and 2011, which included topical experts and stakeholders representing non-governmental organizations, other levels of government, tribes, and the private sector.[4] Often, these individuals were highly experienced in facilitating processes or conducting different kinds of analyses. However, merely inviting a broad range of people to workshops was insufficient to guarantee a diverse set of participants at a given workshop. Instead, stakeholders needed to understand the value of their participation – both for themselves and for the assessment process – before they would participate (NRC 2007a). The experience of recruiting attendees for these workshops and the comments of attendees during the workshops resulted in improved communication about the benefits of participation and the value of meaningful stakeholder engagement to participants and NCADAC (e.g., USGCRP 2010b).

Another early attempt to provide an opportunity for stakeholders to engage in the assessment process was a September 2010 call for public comments on the objectives, proposed topics, and next steps for the assessment (75 FR 54403). The call resulted in input from 25 commenters, including individuals, organizations, and other groups (e.g., a college seminar class) (USGCRP 2011a). The comments were used by the USGCRP and the NCADAC in further developing the assessment process, topics, and structure.

A formal engagement strategy The NCA engagement strategy, approved by the NCADAC in May 2011, built on these early public engagement efforts and laid out a coordinated approach to public participation and communication (NCADAC 2011b). As this guiding document explained,

> "[t]he goal of engaging a broad range of stakeholders in the NCA (as with similar environmental assessment and decision-making processes at all levels of government) is

[4] All workshops described and outputs archived at http://www.globalchange.gov/engage/process-products/NCA3/workshops.

to create a more effective and successful NCA – *improving the processes and products of this effort so that they are credible, salient, and legitimate and build the capacity of participants to engage in the creation and use of these processes and products for decision-making*." (p.3, emphasis added).

The strategy encouraged conversations and learning throughout the assessment process, laid out numerous engagement opportunities for stakeholders throughout the creation of the assessment report, and ensured transparency, with the aim of increasing use of assessment findings in decision-making.

The strategy also called for leveraging existing capabilities inside and outside the Federal government to communicate about NCA3 and create a variety of opportunities for public participation in the assessment process. Recognizing that the success of engagement efforts would rely on building collaborative relationships with individuals and organizations with existing connections to the broader stakeholder communities which NCA3 was targeting, the strategy proposed the development of a cross-sectoral "network of networks" that would serve as a place to share information and co-produce knowledge and engagement efforts. Implementation of these efforts through NCAnet is described below.

Specific plans were also developed and implemented for the release of the public review draft (January 2013) and the final report (May 2014). Importantly, the public review draft was released under the purview of the NCADAC, whereas the final NCA3 was released as a major report of the US government. This required full approval and acceptance from all levels of the Federal government following a legally defined review process and close collaboration and coordination with the White House on the actual release of the NCA3. While acceptance of the report by the Administration was never in doubt, the full nature and format of the release was not assured until days before the intended release date, requiring substantial contingency planning independent of the White House.

4 Implementing engagement in the Third National Climate Assessment

Stakeholder engagement throughout the NCA3, as mentioned, involved opportunities for communication and participation, i.e., : enabling NCADAC and authors to communicate with and engage stakeholders effectively, while creating repeated opportunities for diverse sets of stakeholders to learn about, participate in, provide input into, and communicate about the assessment. This duality is apparent throughout the three engagement phases described below; a list of specific engagement activities in each of the phases is provided in Supplementary Material 2.

4.1 Engagement during the development of NCA3

Request for information and participation One of the innovative mechanisms used in the NCA3 was a request for information that invited contributions of technical inputs or other capacity related to regional, sectoral, and cross-cutting topics proposed for the NCA report and the ongoing NCA process (76 FR 41217). The NCADAC Engagement Working Group and the NCA Office provided descriptions of the potential technical inputs (e.g., literature reviews, case studies, topical reports) and capacities (e.g., hosting events, trainings, activating their networks to participate in various activities) and suggested best practices for developing inputs using open, transparent, and participatory processes (USGCRP 2011b).

In response, approximately 500 technical inputs were received from approximately 200 sources, representing more than 1000 individuals. Inputs ranged from photographs and short descriptions of local impacts of climate change, to previously published papers, to novel scientific work by teams of experts. NCA staff reviewed and catalogued all inputs and offered them for review to NCA3 author teams regarding relevance, topical and technical appropriateness, and adherence to information quality standards.[5]

Suggestions and guidance on engagement Although best practices for engagement were provided along with the initial request for information, some technical input providers were able to implement these suggestions better than others. Often the ability to implement stakeholder engagement was contingent on external monetary resources to support activities and staff time. For example, the technical input to the NCA3 for the US Southwest region used resources available through NOAA's Regional Integrated Sciences and Assessments program to conduct a large workshop and three teleconferences to gather information from stakeholders about the issues they wanted addressed in the report (Garfin and Jardine 2013). Similarly, the US Department of Agriculture sponsored a workshop that focused on the impacts of climate change on rural communities (Hauser and Jadin 2012). Other teams drew on the results of recent stakeholder engagement efforts that, while not focused specifically on the NCA process, revealed key concerns over climate change impacts on particular regions or sectors.

A network of networks (NCAnet) Prior assessments provided important lessons about the crucial role of communication and engagement in making assessments impactful (NRC 2007a). First, direct involvement in an assessment builds familiarity, trust, and greater legitimacy. Second, salience is enhanced by integrating, early and often, the viewpoints and needs of potential end users. Third, communication and engagement are more effective when done by groups and individuals outside the Federal government with whom stakeholders are more familiar. Thus, a strategic decision was made to develop a network of organizations whose participation and interaction with the NCA process would be enabled and facilitated through the NCA Engagement Coordinator and occur through an accessible interface. This network, called NCAnet, extended the already substantial capacity of experts directly involved in the NCADAC and on author teams and served as an essential mechanism for dialogue between NCA3 and outside stakeholders. This network grew steadily from its founding in 2012, largely through word-of-mouth and direct appeal, and now includes more than 170 organizations linked to hundreds of thousands of stakeholders (Fig. 3). Participants represent a wide range of organizations, including professional societies; local, state, and tribal governments; NGOs; business and industry; and academic institutions. Organizations participate in NCAnet voluntarily and generally without financial support for any of their services to the NCA. They have self-organized into topical affinity groups (e.g., education and communication) and have expressed great appreciation for the opportunity to collaborate and exchange ideas.[6]

NCAnet was essential in the development and rollout of NCA3's draft and final report. During the development of NCA3, NCAnet participants contributed technical inputs, organized and joined regional town hall meetings, informed their members about NCA3, hosted

[5] Author teams were responsible for deciding whether cited source material met information quality standards. NCADAC-developed guidance for assuring information quality is available from http://www.globalchange.gov/sites/globalchange/files/NCADAC-Nov2011-Information-Quality-Principles.pdf.

[6] More information about the composition and operation of NCAnet at http://ncanet.usgcrp.gov.

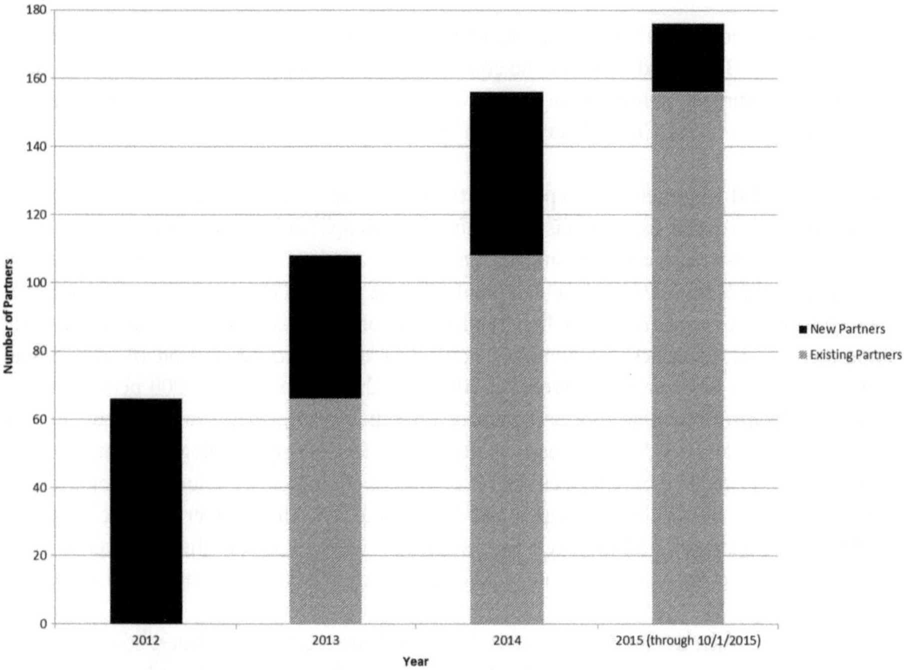

Fig. 3 Growth of NCAnet, January 2012 – October 2015

webinars that sought input on specific topics, and assisted in communication and media outreach during the release of the draft report (an occasion for which the NCADAC and author teams were prepared to respond to inquiries, but did not proactively try to get press coverage, as the product was not yet final). NCAnet was influential in getting significant press coverage for the draft report (approximately 140 media mentions within the first week after the release on January 11, 2013 and approximately 310 mentions by the time comments closed on April 12, 2013; prior to the release of the final NCA3 in May 2014, the draft had been cited or referred to in over 800 media stories).

Importantly, while there was ongoing communication among NCAnet, the NCA Engagement Coordinator, and the NCADAC Engagement Working Group, all activities of the NCAnet were voluntary and completely autonomous. While very substantial trust and cooperative spirit was created over time, there was no control or formal accountability mechanism between NCAnet and the NCADAC or the NCA Office.

Public comments on the draft report The public could comment on the draft report during a 90-day period (January to April 2013). In addition to the Federal Register Notice announcing the comment period (78 FR 4132) and NCAnet participants calling on their own networks to comment, USGCRP agencies sponsored eight public town hall meetings (one per region).[7] NCAnet participants hosted additional meetings and webinars. At these events, report authors provided summaries of draft report findings and attendees could ask questions about the draft report. The meetings often also included sessions focusing on particular sector- or region-

[7] All town halls and related materials available at http://www.globalchange.gov/engage/process-products/NCA3/workshops#Town Halls.

specific topics and on local or regional responses to climate change and provided opportunities for networking. By the end of the public comment period, USGCRP received 4161 comments from 644 government, non-profit, and commercial sector employees, educators, students, and the general public (USGCRP 2014a; Melillo et al. 2014b).

Less successful engagement experiments While many engagement efforts worked exceedingly well, several experiments were limited in scope and impact. One that ultimately proved too labor-intensive for the impact it achieved was the Climate Conversations project, convened by the Keystone Center with a grant from the Energy Foundation. The Center hosted four Climate Conversations across the country, with audiences not already convinced of the importance and seriousness of climate change (for a detailed description of the Climate Conversations, see Moser and Berzonsky, 2015). Of the approximately 2000 people invited, about 200 from different sectors participated. Over half a day, they learned about regional climate changes and NCA3, then participated in World Café-style dialogues about interests and concerns related to those changes. The experiments confirmed the value of dialogue and of educating people about climate change and NCA3, but the effort to recruit participants, the facilitation of dialogue, and the need for meaningful ways to sustain the newly established relationships were not commensurate with the perceived benefits for the NCA (Keystone Center 2012).

Other ideas, such as building a corps of "NCA Ambassadors" – trusted communicators who could convey the findings of the NCA3 once the final report was released – did not come to fruition as they overlapped with existing or new efforts organized outside of the NCA3 process.

Science translation To ensure the report used understandable language and graphics, NCA3 employed an editorial team that assisted the authors with writing in an accessible manner and producing easily comprehensible illustrations. Public comments on the draft report also pointed to areas where readers were confused about concepts and terminology, allowing authors and editors to make further improvements in writing style.

4.2 Engagement during the rollout of NCA3

Engagement workshop In February 2014, in preparation of the release of the final NCA3 report, the NCA Engagement Coordinator organized a workshop for approximately 80 NCAnet participants, NCADAC members, Federal agency representatives, and chapter authors. The workshop sought to generate ideas for engaging assessment users in learning and conversation about climate change impacts and science, using the NCA3 as a springboard. Participants developed ideas and planned implementation of such activities around the release of NCA3, including media outreach, regional events, professional development training, and educational materials (NCAnet 2014, NOAA et al. 2014). Several restrictions affected but did not curtail these activities: resources for NCAnet-led activities had to come from NCAnet organizations; Federal agencies had to work within their own budgets and capacities; and travel funding for outreach was available only for authors and NCADAC members. Planning for these activities continued throughout spring 2014 and following the release of NCA3.

Communication training for authors and NCADAC members: Prior to the release of NCA3, the NCADAC Engagement Working Group and NCA3 editorial team offered several web-based communication trainings for interested authors and NCADAC members. These web trainings allowed authors to practice delivery of report findings and answering questions. In addition, the strategic communications firm Climate Nexus, an NCAnet participant, led an in-person, two-day media training for approximately 20 NCADAC members and NCA3 authors. Although several of these authors were already media-experienced, the training session was particularly useful for helping individuals sharpen their message and communicate the NCA3 findings more effectively.

Communication products One of the innovations of NCA3 was its electronic delivery. Only short report summaries (a 140-page Highlights document, a 20-page Overview, and 4-page regional summaries) were printed; the vast body of material of the assessment was delivered via an interactive website (http://nca2014.globalchange.gov). The website was designed to work on virtually any desktop or mobile platform and to easily connect to social media (Facebook, Twitter, and permalink sharing options for each chapter segment, graphic, and key finding). In addition to the direct derivatives of the NCA3 report, one of the NCA3 editors independently produced short video introductions to selected chapters, featuring the chapters' lead authors, and to selected topics, featuring Americans dealing with particular impacts in their regions or sectors (https://vimeo.com/channels/nca). These videos were featured repeatedly in the TV and online media coverage.

Release day events The NCA3 was required by Congress and prepared by an advisory committee; when accepted by the Obama Administration, it became a government report released to Congress and the public. Following the commitment of the Administration to advance climate policy during the President's second term, the White House devoted time and personnel to creating a highly visible release event on May 6, 2014. After the NCADAC approved the document and NOAA delivered it to the White House, Administration leaders together with the chair of the NCADAC held a press conference to announce the key findings of the assessment. Later that day, the President hosted several one-on-one interviews with selected TV weathercasters in the White House Rose Garden. Concurrently, the President's science advisor and other Administration officials hosted a live-streamed event for stakeholders, during which a dozen NCA3 authors introduced key findings from their chapters.[8]

Outreach immediate following the report release In the first few days following the release and after White House involvement subsided, NCAnet participants, the NCADAC, and author teams continued a substantial schedule of outreach activities, including a briefing to Congress the day after the release. Over 100 activities (ranging from web-based seminars to community meetings to full-day workshops) have taken place since the release, many of them organized by NCAnet participants.

4.3 Engagement in support of the sustained assessment

Implementation of the core elements of the sustained assessment process as advised by the NCADAC (Buizer et al. 2013; Buizer et al. 2015) has been uneven. However, selected

[8] The stakeholder event is archived at http://www.c-span.org/video/?319224-2/white-house-unveils-climate-assessment-report.

elements continue, including requests for information (80 FR 26105) and Federal Advisory Committee nominations (80 FR 45643), special assessment reports, launch of a pilot indicator system (Kenney and Janetos 2015), and NCAnet. Two of these are described below.

Ongoing NCAnet activities Congruent with the intention to build a sustained assessment process (Buizer et al. 2015), NCAnet has persisted. According to a brief informal survey of NCAnet participants conducted in spring 2015, considerable outreach work around NCA3 continues and many participants have developed regional or topical reports that draw on information from NCA3. Several of the affinity groups continue to meet to create and refine products and activities, including on the anniversary of the report release. In addition, new affinity groups have formed to address emerging topics such as valuation, risk management, and climate projections.

Climate and health assessment One of the special reports currently in development is an assessment on climate change and human health, led by USGCRP's Climate Change and Human Health Working Group. The topic was chosen because of significant external stakeholder and Federal agency interest, and was mentioned in the President's Climate Action Plan. As during NCA3, this special report requested public input to inform the report's scope (79 FR 7417) and during a public comment period for the draft assessment (80 FR 18619).

5 Preliminary evaluation of impact

The overall goal of engagement was to create a more effective and successful NCA3 – i.e., an assessment that is viewed by participants and outsiders as credible, salient, and legitimate, underlain by a transparent and accessible process (NCADAC 2011b, NRC 2007b). The NCA engagement strategy suggested that NCA3 could achieve an even broader impact, namely, "[t]he NCA process and products…can serve as a vehicle for civic engagement, providing space for conversations about the underlying science, expected impacts of, and responses to climate change in the US" (NCADAC 2011b, p. 3).

Soon after the NCA3 release, USGCRP organized a workshop on how to conduct a critical but constructive post-NCA3 evaluation and how to build ongoing evaluation efforts into the sustained assessment process (USGCRP 2014b). The workshop involved evaluation experts from inside and outside government, academia, and NGOs. While the NCADAC had built careful tracking and ongoing learning-oriented evaluation into developing NCA3, including its engagement dimension, a full external evaluation has not yet been undertaken.

Selected elements of the engagement process, however, have been critically and routinely assessed. For example:

- Regional town halls and workshops included participant feedback mechanisms;
- The Climate Conversations were critically debriefed by the dialogue facilitators, funders, NCADAC Engagement Working Group, NCA Engagement Coordinator, and NCA leadership;
- A media analysis was conducted after the release of the draft NCA3 report to track responses to the report and process;
- Ongoing media and outreach activities tracking is being undertaken by USGCRP and NCAnet members;

- Researchers from Michigan State University, together with USGCRP staff, conducted social network analysis (SNA) to evaluate how NCA3 outreach activities have changed the interconnectivity among researchers, government actors, NGOs, and other stakeholders (Supplementary Material 3); and
- NCAnet members provide regular feedback about their use of NCA3 and usefulness of the NCAnet process as a part of NCAnet conversations.

Based solely on this ongoing internal tracking, it is impossible to assess the overall impact of the assessment. Thus, a critical evaluation must be left to external experts not involved in the day-to-day operation of NCA3.

In the absence of an independently conducted evaluation, however, ongoing tracking of the assessment's uptake via surveys and interviews provides hints of the assessment's perceived credibility, salience, legitimacy, and transparency. For example, the assessment has been cited as the rationale and policy justification for several Executive Orders and Federal agency climate adaptation initiatives (e.g., EO 13514, the President's Climate Action Plan, EO 13653, EO 13693), though some representatives of the US Congress have tried to prevent policy initiatives based on NCA3. On subnational scales, NCA3 and its underlying regional technical input reports have informed regional and state-level efforts (e.g., Bathke et al. 2014; Allegheny Highlands Climate Change Impacts Initiative 2015; Tassel 2015). NGOs and private sector actors have also drawn on the findings of NCA3 (e.g., Adams et al. 2014; Risky Business 2014).

As for reaching the American public, the widespread media coverage gives some hope. In recent years, communication about climate change – relatively muted in traditional media (Boykoff 2015) but ongoing, loud, and often polarizing in new and social media – has struggled to engage the American public in constructive discourse (Moser and Berzonsky 2015). Previous NCA reports and other scientific assessments were not paired with comprehensive communication and outreach plans (Ekwurzel et al. 2011). By contrast, active White House engagement, involving the President and other Administration officials, and interviews with weathercasters who are among America's most trusted climate change messengers (Maibach et al. 2011; Supplementary Material 4), live-streamed release events, and the enormous outreach efforts undertaken by the NCAnet member organizations likely contributed to the "news splash" when NCA3 was released and thereafter. Within one week of the release, more than 2000 news stories citing NCA3 were cataloged; media mentions have continued at a steady rate (over 5000 unique news stories cited NCA3 by October 2015, often 5 to 10 per week). Within the first year of its release, the full report was downloaded over 850,000 times, the report Highlights were downloaded over 189,000 times, and over 433,000 users visited the NCA3 website.

6 Conclusions: lessons for the sustained assessment

The NCA3 engagement efforts can teach several important lessons for future national and other assessments.

Making engagement a priority Despite early agreement by USGCRP's participating Federal agencies and the NCADAC that engagement and communication would be priorities for NCA3, in practice science was often treated preferentially. For example, some technical input teams did not engage potential information users, the Federal assessment plan lacked a dedicated budget for engagement, and there was tension, though diligently handled, between authors and editors regarding the creation of succinct and accessible key messages and chapters. While the engagement strategy approved by the NCADAC provided a framework to ensure engagement would be a priority, many ideas could not be implemented due to a lack of staff time and resources.

Building extended stakeholder networks One of the core pillars of the sustained assessment process is the importance of building and maintaining collaborations with an extended community of scientists and others reaching into stakeholder communities. NCA3 has had unprecedented success in this effort. As the SNA (Supplemental Material 3) showed, the network of stakeholders and extent of interactions between people and organizations engaged by NCA3 has grown significantly over time. As a tool in support of ongoing engagement efforts, SNA can also point to network gaps and guide outreach efforts to bring additional organizations into NCAnet for the sustained assessment.

Matching assessment and engagement boundaries Boundaries of the regions delineated in the report were adjusted slightly from those used in NCA1 and NCA2, to align with state boundaries, allow stakeholders to more quickly be identified or locate themselves within the assessment, and for information to align with common jurisdictional boundaries. In addition, the "Islands" category used previously was changed, placing Puerto Rico and the U.S. Virgin Islands in the Southeast and Hawai'i and the Pacific Islands in their own region. However, the placement of some states within particular regions and the size of some of the regions proved to be confusing for some and may require further deliberation in future assessments.

Stakeholder-driven topical coverage Several new topics were added to NCA3 on the basis of public and NCADAC input. Most notable were multiple calls for cross-sectoral chapters (e.g., Energy, Water, Land; Land Use and Land Cover Change; and Rural Communities) and response chapters (Decision Support; Adaptation; and Mitigation). These are of growing importance and continued stakeholder engagement will ensure improved decision-relevance. Undoubtedly, stakeholder input is important to capture emerging information needs.

Clear, accessible language and visuals, electronically delivered The linguistic, electronic, and visual access to the often dense and complex climate change information was crucial, and countless comments from outsiders reinforced the importance of providing information in this way.

Collaboration, not outsourcing Rather than counting on individuals to write a specific section of a chapter or to create and lead one portion of an engagement activity, NCA3 experience indicates that the most useful input is developed collaboratively. The requisite skills for transdisciplinary work should continue to be built and fostered.

Dedicated and sustained resource stream is essential to engagement process Having dedicated staff for the engagement and an expertise-rich NCADAC were essential to the success of NCA3. Building the sustained assessment without at least this level of support is likely to result in inadequate outcomes. Several engagement ideas could only be accomplished with resources, staff time, and financial support from NCAnet members and other external funding sources. While an indication of great generosity and dedication, this is not a sustainable model for engaging the American public over the long haul. Effective engagement – given its central importance in reaching those who could actually use the results of considerable Federal investment in science – is a necessary, not supplementary, investment in America's future preparedness for climate change.

References

Adams P, Steeves J, Ashe B, Firth J, Rabb B (2014) Climate risks study for telecommunications and data center services, Report prepared for the General Services Administration. https://sftool.gov/Content/attachments/GSA%20Climate%20Risks%20Study%20for%20Telecommunications%20and%20Data%20Center%20Services%20-%20FINAL%20October%202014.pdf. Accessed 10 May 2015

Allegheny Highlands Climate Change Impacts Initiative (2015) On the chopping block: the impacts of global warming and climate change on the mid-atlantic allegheny highlands. http://www.alleghenyclimate.org/documents/Updated%201-19-15%20OTCB%20Report.pdf. Accessed 10 May 2015

Bathke DJ, Oglesby RJ, Rowe CM, Wilhite DA (2014) Understanding and assessing climate change: Implications for Nebraska. http://snr.unl.edu/download/research/projects/climateimpacts/2014ClimateChange.pdf. Accessed 10 May 2015

Boykoff MT (2015) 2000–2015 United States newspaper coverage of climate change or global warming (Updated Through May 2015). http://sciencepolicy.colorado.edu/media_coverage/us/index.html. Accessed 3 June 2015

Buizer JL, Fleming P, Hays SL, Dow K, Field CB, et al. (2013) Report on preparing the nation for change: Building a sustained National Climate Assessment process. http://downloads.globalchange.gov/nca/NCADAC/NCADAC_Sustained_Assessment_Special_Report_Sept2013.pdf. Accessed 11 May 2015

Buizer JL, Dow K, Black ME, Jacobs KL, Waple A, Moss RH, Moser S, Gustafson DI, Richmond TC, Hays SL, Field CB (2015) Building a sustained climate assessment process. Climatic Change. doi:10.1007/s10584-015-1501-4

CCSP (2003) Strategic plan for the U.S. Climate Change Science Program. http://data.globalchange.gov/assets/2a/42/f55760db8a810e1fba12c67654dc/ccsp-strategic-plan-2003.pdf. Accessed 15 October 2015

Dilling L, Lemos MC (2011) Creating usable science: opportunities and constraints for climate knowledge use and their implications for science policy. Glob Environ Chang 21:680–698

Ekwurzel B, Frumhoff PC, McCarthy JJ (2011) Climate uncertainties and their disconents: increasing the impact of assessments on public understanding of climate risks and choices. Clim Chang 108:791–802

Farrell AE, Jäger J, VanDeever SD (2006) Overview: Understanding design choices. In: Farrell AE, Jäger J (eds) Assessments of regional and global environmental risks: designing processes for the effective use of science in decision making. Resources for the Future, Washington, DC, pp. 1–24

Garfin G, Jardine A (2013) Overview. In: Garfin G, Jardine A, Merideth R, Black M, LeRoy S (eds) Assessment of climate change in the southwest United States. Island Press, Washington, DC, pp. 21–36

Hauser R, Jadin J (2012) Rural communities workshop. Technical report. http://downloads.globalchange.gov/nca/technical_inputs/rural-communities-workshop-technical-input.pdf. Accessed 7 May 2015

IPCC (2014) climate change 2014: Synthesis report. Contribution of working groups I, II and III to the fifth assessment report of the Intergovernmental Panel on Climate Change. IPCC, Geneva

Jäger J, Farrell AE (2006) Improving the practice of environmental assessment. In: Farrell AE, Jäger J (eds) Assessments of regional and global environmental risks: designing processes for the effective use of science in decision making. Resources for the Future, Washington, DC, pp. 278–293

Jacobs KL, Buizer JL (2015) Building community, credibility and knowledge: the third US National Climate Assessment. Climatic Change. doi:10.1007/s10584-015-1445-8

Karl TR, Melillo JM, Peterson TC (eds) (2009) Global climate change impacts in the United States. Cambridge University Press, New York

Keller AC (2010) Credibility and relevance in environmental policy: measuring strategies and performance among science assessment organizations. J Public Adm Res Theory 20:357–386

Kenney MA, Janetos AC (2015) Building an integrated national climate indicator system. Climatic Change (this issue)

Keystone Center (2012) Final report and lessons learned: the climate conversations project. Submitted to The Energy Foundation. The Keystone Center, Washington, DC. https://www.keystone.org/policy-initiatives-center-for-science-a-public-policy/environment/climate-conversations.html. Accessed 3 June 2015

Maibach E, Cobb S, Leiserowitz A, Peters E, Schweizer V, et al. (2011) A national survey of television meteorologists about climate change education. Center for climate change communication. George Mason University, Fairfax VA

McNie EC (2007) Reconciling the supply of scientific information with user demands: an analysis of the problem and review of the literature. Environ Sci Pol 10:17–38

Melillo JM, Richmond TC, Yohe GW (eds) (2014a) Climate change impacts in the United States: the Third National Climate Assessment. U.S. Global Change Research Program, Washington, DC http://nca2014.globalchange.gov/. Accessed 3 June 2015

Melillo JM, Richmond TC, Yohe GW (2014b) Appendix 1: Report development process. In: Melillo JM, Richmond TC, Yohe GW (eds) Climate change impacts in the United States: The Third National Climate Assessment. U.S. Global Change Research Program, Washington, DC, pp. 727–732

Morgan MG, Cantor R, Clark WC, Fisher A, Jacoby HD, et al. (2005) Learning from the U.S. national assessment of climate change impacts. Environmental Science & Technology 39:9023–9032

Moser SC (2005) Stakeholder involvement in the first U.S. national assessment of the potential consequences of climate variability and change: an evaluation, finally. http://www.susannemoser.com/documents/Moser_Draft_2-6-05.pdf. Accessed 6 June 2014

Moser SC, Berzonsky C (2015) There must be more: Communication to close the cultural divide. In: O'Brien K, Selboe E (eds) Culture and the adaptive challenge of climate change. Cambridge University Press, Cambridge, pp. 287–310

Moss RH (2015) Assessing decision support systems and levels of confidence to narrow the climate information "usability gap". Climatic Change. doi:10.1007/s10584-015-1549-1

Moss R, Scarlett PL, Kenney MA, Kunreuther H, Lempert R, et al. (2014) Decision support: Connecting science, risk perception, and decisions. In: Melillo JM, Richmond TC, Yohe GW (eds) Climate change impacts in the United States: The Third National Climate Assessment. U.S. Global Change Research Program, Washington, DC, pp. 620–647

National Research Council (NRC) (2007a) Analysis of global change assessments: lessons learned. National Academies Press, Washington, DC

NOAA, NCAnet Education Affinity Group, and CLEAN Network (2014) National Climate Assessment (NCA) teaching resources. http://wwwclimategov/teaching/2014-national-climate-assessment-resources-educators Accessed 3 June 2015

NRC (2007b) Evaluating progress of the US Climate Change Science Program. National Academies Press, Washington, DC

NRC (2008) Public participation in environmental assessment and decision making. National Academies Press, Washington, DC

NRC (2010a) Advancing the science of climate change. National Academies Press, Washington, DC

NRC (2010b) Informing effective responses to climate change. National Academies Press, Washington, DC

NCADAC (2011a) National Climate Assessment: interim strategy – summary. http://wwwglobalchangegov/sites/globalchange/files/NCADAC-May2011-Interim-Strategypdf Accessed 4 May 2015

NCADAC (2011b) National Climate Assessment (NCA) engagement strategy. http://downloads.globalchange.gov/nca/nca-engagement-strategy_5-20-11.pdf. Accessed 4 May 2015

NCAnet (2014) Workshop on building capacity for engagement around the Third National Climate Assessment Report. http://ncanetusgcrpgov/partners/calendar/nca3-engagement-workshop Accessed 10 May 2015

Parson EA, Corell RW, Barron EJ, Burkett V, Janetos A, et al. (2003) Understanding climatic impacts, vulnerabilities, and adaptation in the United States: building a capacity for assessment. Clim Chang 57:9–42

Risky Business Project (2014) Risky business: the economic risks of climate change in the United States. http://riskybusinessorg/uploads/files/RiskyBusiness_Report_WEB_09_08_14pdf Accessed 10 May 2015

Tassel S (2015) Weather we don't recognize: How climate change is affecting the Midwest's weather and how communities are responding. Island Press, Washington, DC

USGCRP (2010a) Strategic planning workshop. http://dataglobalchangegov/assets/46/90/493c479d7145a7087a0292919c04/strategic-planning-workshop-reportpdf Accessed 15 October 2015

USGCRP (2010b) Planning regional and sectoral assessments for the National Climate Assessment. http://data.globalchange.gov/assets/aa/bb/1bc8653ec0e21745a9a1584a7679/regional-sectoral-workshop-report.pdf. Accessed 15 October 2015

USGCRP (2011a) Comments received. http://downloadsglobalchangegov/nca/nca_frn-sep_comments_compiled_2-7-11-posting-xls Accessed 11 May 2015

USGCRP (2011b) Supplement to: Request for information: technical inputs and assessment capacity on topics related to 2013 U.S. National Climate Assessment. http://www.globalchange.gov/sites/globalchange/files/nca-rfi-supplement_7-11.pdf Accessed 11 May 2015

USGCRP (2014a) Comments received on public draft Third National Climate Assessment. http://wwwglobalchangegov/sites/globalchange/files/NCA3-Public-Commentspdf Accessed 11 May 2015

USGCRP (2014b) Frameworks for evaluating the National Climate Assessment: workshop report. http://downloadsglobalchangegov/nca/workshop-reports/NCA-Evaluation-Frameworks-Workshop-Report-2014-finalpdf Accessed 11 May 2015

USGCRP (2015) First National Climate Assessment 1997–2000. http://wwwglobalchangegov/engage/process-products/NCA1 Accessed 4 May 2015

Climatic Change (2016) 135:55–68
DOI 10.1007/s10584-015-1494-z

Innovations in science and scenarios for assessment

Kenneth E. Kunkel[1,2] · Richard Moss[3] · Adam Parris[4]

Received: 2 February 2015 / Accepted: 16 August 2015 / Published online: 29 August 2015

Abstract Scenarios for the Third National Climate Assessment (NCA3) were produced for physical climate and sea level rise with substantial input from disciplinary and regional experts. These scenarios underwent extensive review and were published as NOAA Technical Reports. For land use/cover and socioeconomic conditions, scenarios already developed by other agencies were specified for use in the NCA3. Efforts to enhance participatory scenario planning as an assessment activity were pursued, but with limited success. Issues and challenges included the timing of availability of scenarios, the need for guidance in use of scenarios, the need for approaches to nest information within multiple scales and sectors, engagement and collaboration of end users in scenario development, and development of integrated scenarios. Future assessments would benefit from an earlier start to scenarios development, the provision of training in addition to guidance documents, new and flexible approaches for nesting information, ongoing engagement and advice from both scientific and end user communities, and the development of consistent and integrated scenarios.

1 Introduction

The US National Climate Assessment's (NCA) need for observations and future scenarios of socioeconomic, climate, and environmental conditions arises from requirements in the 1990

This article is part of a special issue on "The National Climate Assessment: Innovations in Science and Engagement" edited by Katharine Jacobs, Susanne Moser, and James Buizer.

Electronic supplementary material The online version of this article (doi:10.1007/s10584-015-1494-z) contains supplementary material, which is available to authorized users.

✉ Kenneth E. Kunkel
ken.kunkel@noaa.gov

1 Cooperative Institute for Climate and Satellites-North Carolina, North Carolina State University, Asheville, NC, USA

2 NOAA National Centers for Environmental Information, 151 Patton Avenue, Asheville, NC 28801, USA

3 University of Maryland, College Park, MD, USA

4 Climate Program Office, National Oceanic and Atmospheric Administration, Silver Spring, MD, USA

Global Change Research Act (GCRA). Specifically, in preparing and submitting an assessment to the President and Congress, the report should, among other things, "analyze current trends in global change, both human-induced and natural, and project major trends for the subsequent 25 to 100 years" (GCRA, Section 106). Responding to this requirement, the Third NCA (NCA3) undertook an ambitious effort to provide new information on past and potential future conditions for the assessment process. The scenario effort was planned by a working group of the National Climate Assessment Development and Advisory Committee (NCADAC). Implementation drew on the resources and expertise of the US Global Change Research Program (USGCRP), the National Oceanographic and Atmospheric Administration (NOAA), the National Aeronautics and Space Administration, the Environmental Protection Agency (EPA), the Department of Energy (DOE), the Department of the Interior, and other agencies. A technical support unit (TSU) at the National Climatic Data Center prepared climate information, while other components of the scenarios drew on a variety of existing resources published in the peer-reviewed research literature. A key challenge was attempting to move beyond traditional scenarios approaches, focusing instead on facilitating consistency and coordination among those producing the assessment and using scenarios to facilitate communication with end users. Innovations included a set of regional climate outlooks that provided narrative descriptions of past and potential future climate features of importance to assessors and potential end users, a range of sea level change scenarios considering multiple potential users, and resources to support participatory scenario planning as a way to better engage decision makers, resource managers, and communities.

The authors of this paper were all involved in various aspects of the scenario process, so this is not a disinterested evaluation of what was planned and achieved. Rather, our purpose is to describe NCA3 scenario objectives and products, including several innovations from prior NCAs, and to provide our reflections on topics for evaluation and additional innovations for future scenario and assessment planners to consider. We also seek to advance the research literature on preparation and use of multi-component global change scenarios for research and assessment purposes.

We first briefly review the scenarios provided in the first two NCAs, and describe the strategy of the NCADAC's scenario working group (SWG). We then review in detail the climate, sea level, integrated socioeconomic and land cover/land use, and scenario planning components of the scenario strategy. We conclude with questions and issues to be considered for the sustained national assessment and future assessment reports.

2 Scenarios in the 2000 and 2009 NCA reports

The 2000 assessment (NAST 2000) and 2009 report (Karl et al. 2009) were very different in character, which is reflected in their scenarios (Moss et al. 2011). The 2000 NCA was similar to NCA3 in that one of its objectives was to establish networks of assessors and end users that would persist following report completion. Another similarity was its level of ambition: multiple products including multiple regional and sectoral chapters and a range of summary products. Three basic scenarios types were provided: climate (MacCracken et al. 2001); ecosystems/vegetation (Melillo et al. 2001); and socioeconomic (Parson et al. 2001). Selection of climate scenarios was based on availability of data at the time of the assessment. The scenarios were based on two Global Climate Models (GCMs): the U.K. Hadley Model (HadCM2) and the Canadian GCM (CGCM1). Only one emissions scenario was used to 'force' (i.e., as an exogenous input to) the models: the Intergovernmental Panel on Climate Change (IPCC) IS92a (a mid-range 'business as usual' emissions future). Some climate diversity was provided because HadCM2 was relatively cooler and wetter and CGCM1 warmer and drier. Controversy arose because no US models were represented and the use of only one emissions future did nothing to bound the true uncertainty (Morgan et al. 2005). Terrestrial ecosystem/vegetation scenarios were produced through the Vegetation-Ecosystem Modeling and Analysis Project (VEMAP) to provide information on ecosystem shifts resulting from climate change. Socioeconomic projections for a small number of population and economic variables at county scale to 2030 and aggregate national scale projections from 2030 to 2100 were provided from an integrated assessment model. Guidance was provided for assessment authors to develop simple parametric projections of variables of interest for the assessment of impacts in particular places or sectors. According to Morgan et al. (2005), the socioeconomic scenarios were not widely used, and most participants suggested another approach was needed for the future.

By contrast, the 2009 assessment needed to be completed in a short period of time (13 months) to comply with a legal ruling that the "Synthesis and Assessment Products" prepared by the program on a variety of climate science and impacts topics failed to fulfill the requirements of the GCRA. The resulting assessment focused on integrating information contained in these reports, supplemented with more recent findings in some areas. Report authors used climate information based on a set of 15 models from the Coupled-Model Intercomparison Project Phase 3 (CMIP3) forced by the IPCC Special Report on Emissions Scenarios (SRES) B1 (low emissions) and A2 (high emissions) scenarios. Many of these models were statistically downscaled to a higher spatial and temporal resolution using a bias-corrected spatial disaggregation (BCSD) method (Wood et al. 2002). The statistical downscaled data were used to produce several derived climate variables (e.g., number of days> 90 °F). A variety of specialized maps, projections, and indicators were produced using these data and appeared in the report. No socioeconomic scenarios were provided.

3 Overview of NCA3 scenario strategy and innovation

NCA3 scenarios preparation began with workshops in 2010 on scenarios (Moss et al 2011) and modeling (Janetos et al. 2011). These engaged a broad range of researchers, intermediate users (mostly climate modelers and impacts scientists), end users, and USGCRP agency research program managers. The scenarios workshop focused on types of scenarios (climate, socioeconomic, etc.) and their inter-relationships, reviewed previous scenario experience in the NCA and other assessments, provided an overview of ongoing scenario efforts, and included extensive discussion of user needs. Climate, socioeconomic, and user support community discussions were also held to plan scenario material for the NCA3.

Development of a detailed scenario strategy fell to the SWG. The group included approximately 20 NCADAC members and representatives from several federal agencies, incorporating wide research expertise, individuals from the private sector, and users. The group met virtually over a roughly 3-month period and developed recommendations that were presented to the NCADAC in May, 2011 (NCADAC Ad Hoc Working Group on Scenarios 2011) and approved in November 2011. The SWG made five sets of specific recommendations on historical observations and future scenario information for: 1) climate, including traditionally-provided global scenarios, regional downscaling, and narrative climate 'outlooks'; 2) global sea level rise including overview of factors that result in regional anomalies; 3) land cover and land use; 4) socioeconomic conditions; and 5) participatory scenario planning activities. Each of these topics is discussed in greater detail below, but several general points about the SWG's objectives are relevant.

The strategy considered needs for two broad categories of users: scenarios for intermediate users, especially to support and coordinate modeling and synthesis; and scenarios and related tools intended to communicate with lay audiences and to support participatory processes considering the implications of climate change. There is some overlap in these two sets of needs, but there are also important tensions. For example, intermediate users wanted consistent scenarios reflecting scientific consensus on the range of plausible outcomes, whereas the risk managers would be better served by scenarios that take into account low probability, high consequence events (EEA 2009).

The more extensive and diverse set of products envisioned by the SWG was intended to address unmet needs noted by authors of prior assessments (mostly intermediate users) and to provide new types of resources that communicated more effectively with end users. This latter use was seen as supporting the shift to a sustained assessment process based on ongoing interactions between the research and user communities.

One of the most important innovations was the decision to produce 'regional climate outlooks', which included narrative descriptions customized for each region. This approach allowed for comparing and discussing complex scientific results in a more accessible fashion than if only model data sets were provided, thus fulfilling the objective of meeting needs of end users and those intermediate users who looked to scenarios for 'context'. The outlooks provided both historical trends of several climatological variables including means and extremes and discussion of current understanding of the future, based on simulations forced by the A2 (high) and B1 (low) emissions scenarios (NCADAC Ad Hoc Working Group on Scenarios 2011, Appendix 1). These scenarios were selected because they bounded uncertainty about future emissions and socioeconomic conditions and because a large body of impacts research was based on them, facilitating preparation of report chapters.

A second innovation was inclusion of sea level change scenarios for risk framing. Sea level scenarios had not been included in prior assessments, but the increasing visibility of coastal vulnerabilities and need for information about potential future conditions for planning purposes created a strong demand. For risk framing purposes, uncertainty bounds were selected to provide users with high-consequence but scientifically grounded futures. The scenarios covered global mean changes to 2100 and descriptions of the factors that cause regional variations.

The third major advance was preparation of guidelines and materials to support more widespread experimentation with participatory scenario planning. The primary purpose of participatory scenario processes is to identify strategies and make decisions robust to a wide range of future conditions. The idea was to encourage and support lead author teams who expressed interest in using such techniques in workshops and stakeholder engagements associated with preparation of their chapters.

4 Description and innovations in physical climate information

Two groups provided advice on physical climate information: (1) the SWG; and (2) a selected group of climate model experts (CMEs). There was considerable discussion about whether to use CMIP3 or CMIP5 model simulations. Although the CMIP5 archive represented the latest set of coordinated GCM simulations, the primary recommendation was to use the CMIP3 simulations, largely because it was uncertain whether a full set of the CMIP5 simulations could be made available in a timely manner to the NCA3 chapter authors and whether any differences between the simulations would be significant enough to alter key conclusions in the NCA evaluation of impacts. A secondary recommendation was to allow the use of CMIP5 simulations as time and resources permitted, specifically the Representative Concentration Pathways (RCP) 2.6 and 8.5 scenarios, as representing a wider range of outcomes than the CMIP3 scenarios. The low emissions B1 scenario in the SRES family is most comparable to RCP4.5, not the lowest in the RCP family. The lowest, RCP2.6, represents a more stringent mitigation scenario than the previous low-end scenarios and provides a larger decision space for consideration. The NCADAC considered these recommendations and decided to use scenarios based on the CMIP3 A2 and B1 simulations as the primary ones that authors were asked to consider, while also allowing use of CMIP5 simulations for illustrative purposes.

The development of regional outlooks involved about 30 regional experts in addition to climate scientists from NOAA. A common set of information was produced for all of the regions. This information was disseminated in the form of a NOAA NESDIS Technical Report series (NOAA NESDIS TR142; Kunkel et al. 2013a, b, c, d, e, f, g; Stewart et al. 2013; Keener et al. 2013). This series included over 700 pages of material and several hundred graphics on historical trends and future projections. Several of the analyses and graphics documented climate model projection uncertainty.

Many regional experts were involved as technical report authors. This enriched the historical material and insured the inclusion of important elements. Many analytic and graphical depictions were common across the regions, providing a level of uniformity not available in previous assessments. The reports were subjected to an intensive anonymous review process involving about 20 reviewers.

The projection information in TR142 utilized three sources of data—climate model data for 15 CMIP3 models and two downscaled data sets: a statistically-downscaled data set by Hayhoe et al. (2004; 2008) and the dynamically-downscaled data set from the North American Regional Climate Change Assessment Program (NARCCAP 2012). The dynamic (NARCCAP) and statistical downscaled data sets provide better representation of spatial patterns in areas of high topographic variability and, in the case of statistically downscaled data sets, biases are lower because of the bias correction methods used in their development. These higher resolution data sets were used to develop the projection products for derived daily resolution impacts relevant climate variables, such as threshold exceedance number. Because of their advantages over GCM data (better representation of spatial patterns, bias correction), these downscaled data sets were the primary basis for developing projections of metrics of extremes.

The projection information included presentations of both mean changes and model spread as one indicator of model uncertainty. The interpretation and communication of model spread/ uncertainty was an area of focus for the NCA3. For temperature projections, there is a substantial model spread (a factor of 2-3 difference among models), but the multi-model mean warming is very large with respect to historical variations. Thus, unequivocal statements were made about large future warming. Precipitation projections were a very different case because future changes are small and not statistically significant for large portions of the U.S. However, this is largely due to the geographic position of the contiguous U.S., straddling the transition zone between projected wetter conditions at higher mid-latitudes and drier conditions in the subtropical belts. This context was viewed as important in communicating the small projected changes.

About the time that TR142 was published and the public review draft of the NCA3 released, a new statistically-downscaled data set became available (Hayhoe et al. 2013). This data set uses more sophisticated methods for daily downscaling, more suitable for extremes analysis. In particular, the newer method downscales from daily GCM data, thus preserving the daily sequence of weather conditions in the GCM simulation and allowing for sub-monthly temporal patterns outside the range of historical conditions, whereas the older method produces daily resolution data by randomly sampling historical months. Although it was too late to incorporate this newer data set into TR142, all NCA3 figures generated using the older Hayhoe et al. (2004) data set were regenerated using this new data set.

4.1 Reflections and issues for evaluation and planning

As CMIP5 data availability increased, it became clear that a basic analysis of CMIP5 data could be done in time for NCA3, although not in time for any impacts research that would support the work of the regional and sectoral chapter authors. Differences arose within the climate science author team about the emphasis on CMIP3 vs CMIP5, with some authors contending that CMIP3 data were too dated for inclusion, despite the NCADAC decision. A compromise was reached to include a series of graphics in the climate science sections to compare CMIP3 and CMIP5 results.

At the center of this issue was the relatively high level of autonomy of the chapter author teams, combined with their lack of involvement in upfront decisions about content in this critical chapter. The overlap among the advisory groups, author teams, and NCADAC decision-making body was relatively small. Earlier selection of author teams and their upfront involvement in decisions could help avoid such conflicts, though there will always be challenges associated with evolving state of knowledge during assessment processes.

There were requests from authors for historical trend and future scenario information on variables other than temperature and precipitation (e.g., wind and solar energy). This need was recognized from the beginning of scenarios development but the time needed for spin-up of capacity limited the scope of analysis. Future assessments may benefit from a sustained assessment process that maintains capabilities across assessment cycles.

Managing the involvement of a large group of regional experts in developing regional outlooks was time-consuming. However, the quality of this group's contributions made this worthwhile. It ensured that regional issues of importance were not overlooked and that the most relevant regional research and analyses were incorporated. For example, information about Lake Champlain ice cover was added to the Northeast Technical Report (Kunkel et al. 2013a) by regional experts and also incorporated into the NCA3 (Walsh et al. 2014, Appendix 3).

5 Sea level change: a new frontier for the NCA

Evidence for global mean SLR has been increasingly documented in assessments of the Intergovernmental Panel on Climate Change (IPCC), the National Research Council (NRC), and a growing body of peer-reviewed scientific literature (IPCC 2001, 2007; Church and White 2011), although the IPCC reports received extensive criticism about the implications of ice melt (e.g., Rahmstorf 2007; Van den Broeke et al. 2011). The IPCC Fifth Assessment Report also includes projections of global mean SLR and the National Research Council published a report on projected SLR in California, Oregon, and Washington in 2012 (NRC 2012). State and local governments and the US Army Corps of Engineers also are developing or have already developed their own SLR projections for coastal planning, policy, and management from the existing body of scientific literature (Parris et al 2012). However, prior to the third NCA, a similar process had not been done at the national scale in an interagency context.

The NCADAC requested an assessment of the scientific literature on global SLR, primarily to help author teams use a consistent approach to projected impacts from future global SLR. A diverse group of experts in science, engineering, and coastal management convened to identify the scenarios. The group was a mix of members from five different federal agencies, eight different academic institutions, and a regional government agency. Thus, the identification of scenarios and synthesis of scientific literature was grounded in rich and diverse experience and expertise, including national coastal risk reduction measures, national coastal management and flood policy, local coastal management policies and problems, and fundamental science on sea level rise and coastal flooding.

Two dimensions of this work proved to be critical. First, for both practical and scientific reasons, the author team consulted with the NCADAC and decided to identify global, rather than regional, scenarios, while still providing a summary of the literature on important processes leading to regional and local variations of sea level change. The agreed-upon goal was scenarios to bound global conditions, within which assessment authors and regional and local experts could conduct more context specific analyses.

Second, the authors embraced an integrated body of social and behavioral science and risk management practice focused on decision making under uncertainty. Specifically, they examined and identified a broad range of scenarios to support preparedness for a range of future conditions. Each scientifically-rigorous scenario was based on plausible future conditions of the Earth system, and on input from managers regarding various coastal management contexts.

The scenarios are bounded by a high-end estimate of global sea level rise by 2100 (6.6 ft or 2 m) and a low-end estimate (8 in. or 0.2 m), with two intermediate scenarios. Scenarios were based on multiple scientific methods: extension of historically observed trends, projections from global climate models, semi-empirical models, and empirical calculations of hypothetical conditions. An additional decision was made not to assign likelihoods or confidence statements to individual scenarios, but to assess confidence in the full range of possible sea level rise by the end of the century. The result was the statement:

> We have very high confidence (>9 in 10 chance) that global mean sea level will rise at least 0.2 m (8 in.) and no more than 2.0 m (6.6 ft) by 2100

While the middle of this range captured by the intermediate scenarios (1 to 4 ft) may be considered more likely, a narrower range also would be necessarily assigned less confidence because peer-reviewed evidence indicates higher amounts of sea level rise are possible by the end of the century. The wide range was intended to support risk management applications. The lowest, intermediate-low, and intermediate-high can all be logically connected to the A2 or B1 emissions scenarios.

5.1 Reflections and issues for evaluation and planning

While intended for use by NCA authors as a foundational document, report preparation extended well into the development of the NCA draft. Input from the NCADAC and peer reviewers improved the document to the point where federal agencies and stakeholders of the NCA perceived utility in providing it to the public for other applications. To date, one of the most prominent uses of the report have been support for the President's Superstorm Sandy Recovery Task Force and national coastal flood risk policy deliberations.

The decision to base the scenarios on multiple scientific methods complicated the justification for each scenario, but is an important innovation. The climate modeling community emphasizes the importance of using an ensemble approach to hedge against the bias of any one model. It is equally important to consider different types of scientific analyses outside of models. Where estimates from different methods or different forms of evidence converge, we can better assess our confidence in estimates of future change.

The risk-based framing proved a useful scenario planning approach, while also addressing NCADAC guidance regarding use of the A2 and B1 emissions scenarios. The Lowest Scenario (0.2 m) is a linear extrapolation of the observed 20th century trend, which could occur under a B1 or RCP2.6 emissions pathway. These pathways require very near-term implementation of aggressive strategies to reduce greenhouse gas emissions – a point explicitly stated in the report. Some authors felt strongly that extrapolation of observed trends provided a rational baseline for a low-end scenario to be compared to model projections. However, the risk-based framing helped distinguish that the Lowest Scenario should only be used where there is a very high tolerance of risk.

6 Land cover/use and socioeconomic scenarios

6.1 NCA3 products and approach

Land cover influences carbon sequestration, regulates water flows and quality, and affects climate through numerous feedback mechanisms (e.g., release of greenhouse gases from land cover changes such as sudden forest dieback or melting of permafrost). Land-based resources support forestry and agriculture products whose production can be affected by changes in climate. Land cover and use in the United States are influenced by numerous factors including climate, but others as well, such as demand for forest and agricultural products, population growth and migration, urbanization, and environmental regulations. Significant changes have occurred over the past several decades, including urban growth, increases in forested areas, and land area reductions for agricultural production.

Numerous government agencies use a variety of technologies for monitoring ongoing changes in land *cover*, including satellite and in situ measurements. Land *use* patterns are more difficult to monitor, and sources of data include those for land cover, plus statistical data associated with planning, zoning, and other environmental and development policy processes. The wide array of data and products became a challenge for the preparation of land cover and use data and scenarios for NCA3, especially in the context of the NCADAC SWG's lack of information regarding the needs of different chapter author teams and potential end users.

The SWG did not converge on a grand design or integrated approach for land cover information, but instead recommended two existing peer-reviewed sources. For baseline land cover data, the National Land Cover Database (NLCD) 2006 (Fry et al. 2011) allocates land cover into 16 different classes and has been applied consistently across all 50 states and Puerto Rico at a spatial resolution of 30 m. For future scenarios of land cover and use, the EPA Global Change Research Program's Integrated Climate and Land Use Scenarios (ICLUS) Project provided downscaled population and land use scenarios based on global SRES scenarios, including B1 (low) and A2 (high) emissions scenarios (see http://www.epa.gov/global-adaptation/iclus/nca_regions.html). Their consistency with the climate scenarios and information was one of the factors that supported their use. The scenarios present information on housing density and impervious surface cover that are consistent with the SRES storylines (Bierwagen et al. 2010).

6.2 Reflections and issues for evaluation and planning

Given the short timelines and lack of consensus on how to produce land cover/use and socioeconomic scenarios, it was fortunate that the existing NLCD and ICLUS products met the minimum NCADAC requirements. But the information provided was most likely less than adequate given the diverse needs across the wide range of ecological and socioeconomic topics covered in the NCA. For example, socioeconomic vulnerabilities to climate change depend in large part on where people are and where they are migrating (demography), what their livelihoods are (economic patterns and changes), how they regulate business and govern daily activities (institutions), what cultural and social patterns influence behavior, and what adaptation options are available (e.g., technologies, management systems, and social networks). Unlike the climate and sea level scenarios, none of the agencies responsible for more operational activities, such as the US Census Bureau and the Bureau of Economic Analysis, were in a position to follow up on their initial interest in supporting the activity. Thus it fell to a small subset of the SWG, with support from the NCA office, to implement the strategy using existing products.

It is unclear how the land cover/use and socioeconomic scenarios were used, if at all. Contributing factors likely included diversity of needs, less familiarity with application of these scenarios, lack of support, and time pressure on author teams. These speculations notwithstanding, development of a better strategy to support the sustained assessment and preparation of future assessment reports will require evaluation of the use of this information in NCA3 and an analysis of the opportunity cost of providing this kind of information.

7 Participatory scenario planning: exploring integration of climate science and applications

7.1 NCA3's pilot approach

The SWG, responding to increased experience with participatory approaches that mix experts and end users in assessment and decision support processes (NRC 2009; Salter et al. 2010), decided to test opportunities for participatory scenario planning as a component of the NCA3 scenario strategy. The approaches use climate and socioeconomic scenarios to anticipate potential local-scale conditions and impacts, and to explore adaptation (and/or mitigation) options that could address potential impacts and withstand the range of potential future conditions.

A SWG subgroup prepared a short guidance document describing the "scenario planning opportunity for regions and sectors" (Hartmann et al. 2012). The strategy requested that those participating in the NCA, including authors and members of the public who were providing inputs to the process or participating as end users, to: 1) inventory and report on scenario planning activities, including identifying and describing groups using scenario planning; 2) incorporate results of ongoing scenario planning activities into chapters; and 3) undertake a pilot scenario planning activity by working with stakeholders and using a scenario method to conduct the adaptation planning. The guidance document for pilot scenario planning activities encouraged stakeholders to explore adaptation options that address potential risks to natural resources, the economy, or their quality of life. It also asked the participants to use projections of potential impacts for their location or sector associated with the B1 and A2 scenarios, drawing on input from the expert science and assessment community. Given these impacts, the next step was to develop narrative scenarios (and if possible, associated quantitative information) on adaptation to these impacts, including specifying needed technologies, financing, institutional developments, and other factors affecting their use. The adaptation scenarios were then to be used as the basis for discussion about the opportunities for implementation, risks, and likely courses of action.

7.2 Reflections and issues for evaluation and planning

None of the chapter author teams took up the specific planning scenario suggestions, though the Southwest Regional Climate Assessment Team did document capacity building activities, including scenario planning activities. Unfortunately there were no resources available for widespread dissemination of the scenario planning guidance, or for providing facilitation for those groups that lacked prior experience with participatory scenario techniques. In the rush to complete the report in a timely manner, "the art of the possible" became the rallying cry for the Assessment team, and this particular request did not seem possible in light of imminent deadlines to produce the NCA3 report. This is another issue that will require evaluation, planning, and adequate resources for future assessments.

8 Conclusions and future recommendations

Ambitious plans for scenarios resulted in many innovations. We offer several final reflections, evaluation needs, and opportunities that should be considered as part of the sustained assessment process and for future assessment reports.

1. Start earlier: Despite best efforts, the scenario materials developed for NCA3 were delivered to authors piecemeal and late. This made it difficult for them to incorporate in their work. Even in the context of a sustained assessment process, there will be an ongoing challenge because the cycle of production of new climate model simulations and impacts literature based on these simulations does not align with scenario needs for the NCA report requirements. For example, the CMIP activities are typically conducted on the approximate 6-year cycle of the IPCC, which is not synchronized with the NCA 4-year cycle. The decision to use CMIP3 produced controversy since CMIP5 was being produced at the same time as the NCA3, and several participants felt the old data were built on old scenarios and models and hence not representative of the state of science. Others argued the differences across CMIP cycles were not significant enough to be an issue for use in impact assessment. Paradoxically, the earlier the start, the greater the potential there is to make this aspect of the challenge more problematic for the cyclical assessment process. Given the need for quality control and vetting with users, work on a new cycle would need to begin before the

end of an ongoing assessment cycle if results were to be available in time. The analytical capabilities that would be available in a sustained assessment process would not (obviously) address the fundamental structural issues of timing, but they would facilitate much quicker incorporation of new information and data sets.

2. Provide training, not just guidance: A recommendation from evaluations of past NCAs (e.g., Morgan et al. 2005) was to provide guidance and training for assessors. For NCA3, detailed guidance documents were prepared for author teams, and webinars were held to provide training on various topics, including scenarios. A web page of scenario resources was also prepared (http://scenarios.globalchange.gov). However, based on anecdotal evidence, the guidance documents did not prove sufficient for most users. A particular issue was that they did not address relationships across topics, for example the need for an integrated approach to risk, uncertainty, and scenarios (see Moss, this issue). Additional training opportunities should be provided for both intermediate and end users. For intermediate users, this should include methodologically-oriented knowledge-sharing workshops and mentoring or coaching to illustrate the proper uses and limits of different approaches, including increasing emphasis on the use of socioeconomic scenarios, scenario planning, and risk framing approaches.
3. Develop new, flexible approaches for nesting information within multiple scales and sectors: NCA3 again reverted to fixed, regional boundaries to inform regional chapters rather than to be truly useful for decisions at multiple scales. Although difficult, it should be possible to enable users to define the geographic scope, as already exists in several climate change portals. Focusing exclusively on the large and heterogeneous NCA regions missed opportunities for assessing different spatial and institution scales and for addressing sectoral needs. If the Assessment can consistently "nest" information within various national, regional and local scales, the exact boundaries of the regions become much less important (See NRC 2007, Evaluation of Global Change Assessments, which recommends a "nested" approach). A coordinated effort to truly integrate (not just link to) existing agency portals with current NCA resources may be more cost-effective than building a system from scratch, but it is not trivial. Such a system would further scenarios development that can be adopted and tailored at finer scales with contextual details that illuminate key issues, systems dynamics, and tradeoffs for interested communities.
4. Provide ongoing engagement and advice from both the scientific and end user communities in scenario development: Schedules and requirements within agencies supporting the NCA, as well as for lead authors and others likely to be involved in preparing future reports, require that preparations for the next assessment begin as soon as the last one is complete. It is essential to engage the communities involved in the sustained assessment and specific reports to ensure that past efforts are evaluated, needs are clearly specified, and the potential for innovations explored. Federal agencies should assemble an informal community of scenario practice or a more formal advisory body to serve as a sounding board and to develop recommendations on how to best provide scenarios and other information.
5. Ensure consistency of scenarios: A further challenge is capacity for combining domain-specific scenarios (e.g., climate change, sea level rise, population/demographics, land use and land cover change, technology evolution, etc.) into integrated scenarios that are consistent and plausible across domains. An interagency group (Scenarios and Interpretive Science Coordinating Group) within the USGCRP has undertaken efforts to improve scenarios with intentions to: 1) advance collaborative science on critical gaps; 2) enhance

methodologies for use-inspired scenario development, risk framing, and contextual interpretation; 3) develop the next generation scenario work products for model intercomparisons, assessments, and analyses; and 4) improve interagency communications, coordination, and accessibility to knowledge, work products, and technical resources.

The innovations in scenario information for the NCA3 were intended to provide the foundation for a rich and cutting-edge assessment. Not all expectations could be met, however, and the process highlighted other needs to be addressed in the future. A sustained assessment process that evaluates past experience and addresses areas of unmet need will improve scenarios for future assessments and other applications.

Acknowledgments This work was partially supported by NOAA through the Cooperative Institute for Climate and Satellites — North Carolina under Cooperative Agreement NA14NES432003 and by the NOAA Climate Program Office, Climate Observations and Monitoring Program. The DOE Office of Science supports basic research on scenarios and supported several workshops that reviewed the state of science in scenarios and related areas of modeling. Robert Vallario provided very helpful insights on these and related aspects of the NCA3 scenarios development process. NASA provided support for the NCA scenario development process.

References

Bierwagen BG, Theobald DM, Pyke CR, Choate A, Groth P, Thomas JV, Morefield P (2010) National housing and impervious surface scenarios for integrated climate impact assessments. Proc Natl Acad Sci 107:20887–20892

Church JA, White NJ (2011) Sea-level rise from the late 19th to the early 21st Century. Surv Geophys 32:585–602. doi:10.1007/s10712-011-9119-1

European Environment Agency (EEA) (2009) Looking back on looking forward: a review of evaluative scenario literature. European Environment Agency, Copenhagen

Fry J, Xian G, Jin S, Dewitz J, Homer C, Yang L, Barnes C, Herold N, Wickham J (2011) Completion of the 2006 national land cover database for the conterminous United States. Photogramm Eng Remote Sensing 77:858–864

Hartmann H, Jacobs K, Moss RH (2012) National climate assessment: scenario planning opportunity for regions and sectors. Ver: 2012.1.15. Unpublished guidance document distributed through National Climate Assessment Office. See http://www.nesdis.noaa.gov/NCADAC/pdf/nov_16/NCADAC_Mtg_Pres_Nov11_All_Final_111611_5a.pdf for a first draft

Hayhoe K et al (2004) Emission pathways, climate change, and impacts on California. Proc Natl Acad Sci U S A 101:12422–12427

Hayhoe K, et al (2013) Development and dissemination of a high-resolution national climate change dataset, 497 ppp. http://cida.usgs.gov/thredds/fileServer/dcp/files/Hayhoe_USGS_downscaled_database_final_report.pdf

IPCC (2001) Climate change (2001) The scientific basis. Contribution of working group 1 to the third assessment report of the intergovernmental panel on climate change. In: Houghton JT, Ding Y, Griggs DJ, Noguer M, van der Linden PJ, Dai X, Maskell K, Johnson CA (eds) Cambridge University Press, Cambridge, UK, New York, USA

IPCC (2007) Climate change (2007) The physical science basis. Contribution of working group i to the fourth assessment report of the intergovernmental panel on climate change. In: Solomon S, Qin D, Manning M, Chen Z, Marquis M, Averyt KB, Tignor M, Miller HL (eds) Cambridge University Press, Cambridge, UK, New York, NY, USA

Janetos AC, Collins W, Wuebbles D, Diffenbaugh N, Hayhoe K, Hibbard K, and Hurtt G (2011) Climate change modeling and downscaling: issues and methodological perspectives for the U.S. national climate assessment. NCA Report Series, vol. 7. U.S. Global Change Research Program, pp 49

Karl TR, Melillo JT, Peterson TC (Eds) (2009) Global climate change impacts in the United States. Cambridge University Press, pp 189. [Available online at http://downloads.globalchange. gov/usimpacts/pdfs/climate-impacts-report.pdf]

Keener VW, Hamilton K, Izuka SK, Kunkel KE, Stevens LE, Sun L (2013) Regional climate trends and scenarios for the U.S. national climate assessment. Part 8. Climate of the Pacific Islands. NOAA Technical Report NESDIS 142-8, pp 45

Kunkel KE, Stevens LE, Stevens SE, Sun L, Janssen E, Wuebbles D, Rennells J, DeGaetano A, Dobson JG (2013a) Regional climate trends and scenarios for the U.S. national climate assessment. Part 1. Climate of the Northeast U.S. NOAA Technical Report NESDIS 142-1, pp 80

Kunkel KE, Stevens LE et al (2013b) Regional climate trends and scenarios for the U.S. national climate assessment. Part 2. Climate of the Southeast U.S. NOAA Technical Report NESDIS 142-2, pp 95

Kunkel KE, Stevens LE, Stevens SE, Sun L, Janssen E, Wuebbles D, Hilberg SD, Timlin MS, Stoecker L, Westcott NE, Dobson JG (2013c) Regional climate trends and scenarios for the U.S. national climate assessment. Part 3. Climate of the Midwest U.S. NOAA Technical Report NESDIS 142-3, pp 96

Kunkel KE, Stevens LE et al (2013d) Regional climate trends and scenarios for the U.S. national climate assessment. Part 4. Climate of the U.S. Great Plains. NOAA Technical Report NESDIS 142-4, pp 83

Kunkel KE, Stevens LE, Stevens SE, Sun L, Janssen E, Wuebbles D, Redmond KT, Dobson JG (2013e) Regional climate trends and scenarios for the U.S. national climate assessment. Part 5. Climate of the Southwest U.S. NOAA Technical Report NESDIS 142-5, pp 79

Kunkel KE, Stevens LE, Stevens SE, Sun L, Janssen E, Wuebbles D, Redmond KT, Dobson JG (2013f) Regional climate trends and scenarios for the U.S. national climate assessment. Part 6. Climate of the Northwest U.S. NOAA Technical Report NESDIS 142-6, pp 76

Kunkel KE, Stevens LE, Stevens SE, Sun L, Janssen E, Wuebbles D, Dobson JG (2013g) Regional climate trends and scenarios for the U.S. national climate assessment. Part 9. Climate of the Contiguous United States. NOAA Technical Report NESDIS 142-9, pp 78

MacCracken M, Barron E, Easterling D, Felzer B, Karl T (2001) Scenarios for climate variability and change. Pages 13-72 in NAST. Climate Change Impacts on the United States. Cambridge University Press, Cambridge, U.K.

Melillo J, Janetos A, Schimel D, Kittel T (2001) Vegetation and biogeochemical scenarios. In: NAST (ed) Climate change impacts on the United States: the potential consequences of climate variability and change. Cambridge University Press, Cambridge

Morgan MG, Cantor R, Clark WC, Fisher A, Jacoby HD, Janetos AC, Kinzig AP, Melillo J, Street RB, Wilbanks TJ (2005) Learning from the U.S. national assessment of climate change impacts. Environ Sci Technol 39: 9023–9032

Moss RH, Mearns LO et al (2011) Scenarios for research and assessment of our climate future: issues and methodological perspectives for the U.S. national climate assessment. NCA Report Series, Volume 6. US Global Change Research Program, Washington, DC. Available at: http://assessment.globalchange.gov

NARCCAP, cited 2012: North American Regional Climate Change Assessment Program. [Available online at http://www.narccap.ucar.edu/

National Assessment Synthesis Team (2000) Climate change impacts on the United States: the potential consequences of climate variability and change, report for the US global change research program, 163 pp., U.S. Global Climate Research Program, Washington, DC. Available online at: http://library. globalchange.gov/downloads/download.php?id=124

National Climate Assessment Development and Advisory Committee, Ad Hoc Working Group 3, Scenarios and Regional Summaries (2011) Report for NCADAC Meeting May 20, 2011. 37 pp. Available at: http://www. nesdis.noaa.gov/NCADAC/pdf/20b.pdf

National Research Council (2012) Sea-Level Rise for the Coasts of California, Oregon, and Washington: Past, Present, and Future. The National Academies Press, Washington

National Research Council (NRC) (2007) Analysis of global change assessments: lessons learned. National Research Council, Committee on Analysis of Global Change Assessments, Board on Atmospheric Sciences and Climate, Division on Earth and Life Studies. National Academies Press, pp 196

National Research Council (NRC) (2009) Informing decisions in a changing climate, Panel on strategies and methods for climate-related decision support. The National Academies Press, Washington

Parris AP, Bromirski P, Burkett V, Cayan D, Culver M, Hall J, Horton R, Knuuti K, Moss R, Obeysekera J, Sallenger A, Weiss J (2012) Global sea level rise scenarios for the US national climate assessment. NOAA Tech Memo OAR CPO-1. pp 37

Parson EA, Morgan MG, Janetos A, Joyce L, Miller B, Richels R, Wilbanks TJ (2001) The socioeconomic context for climate impact assessment.In: NAST (ed) Climate change impacts on the united states: the potential consequences of climate variability and change, pp 93–107. Cambridge University Press, Cambridge, U.K.

Rahmstorf S (2007) A semi-empirical approach to projecting future sea-level rise. Science 315:368–370. doi:10.1126/science.1135456

Salter J, Robinson J, Wiek A (2010) Participatory methods of integrated assessment—a review. Wiley Interdiscip Rev Clim Chang 1:697–717

Stewart BC, Kunkel KE, Stevens LE, Sun L, Walsh JE (2013) Regional climate trends and scenarios for the U.S. national climate assessment. Part 7. Climate of Alaska. NOAA Technical Report NESDIS 142-7, pp 61

Van den Broeke MR, Bamber J, Lanaerts J, Rignot E (2011) Ice sheets and sea level: thinking outside the box. Surv Geophys 32:495–505. doi:10.1007/s10712-011-9137-Z

Walsh J et al (2014) Appendix 3: climate science supplement. Climate change impacts in the United States: the third national climate assessment. In: Melillo JM, Richmond Terese TC, Yohe GW (eds) U.S. Global Change Research Program, 735–789. doi10.7930/J0KS6PHH

Wood A, Maurer E, Kumar A, Lettenmaier D (2002) Long-range experimental hydrologic forecasting for the eastern United States. J Geophys Res 107 Art. No. 4429

Climatic Change (2016) 135:69–83
DOI 10.1007/s10584-015-1588-7

Innovations in information management and access for assessments

Anne M. Waple[1] · Sarah M. Champion[2] · Kenneth E. Kunkel[2] · Curt Tilmes[3]

Received: 31 March 2015 / Accepted: 11 December 2015 / Published online: 17 February 2016

Abstract The third National Climate Assessment (NCA3) included goals for becoming a more timely, inclusive, rigorous, and sustained process, and for serving a wider variety of decision makers. In order to accomplish these goals, it was necessary to deliberately design an information management strategy that could serve multiple stakeholders and manage different types of information - from highly mature government-supported climate science data, to isolated practitioner-generated case study information - and to do so in ways that are consistent and appropriate for a highly influential assessment. Meeting the information management challenge for NCA3 meant balancing relevance and authority, complexity and accessibility, inclusivity and rigor. Increasing traceability of data behind figures and graphics, designing a public-facing website, managing hundreds of technical inputs to the NCA, and producing guidance for over 300 participants on meeting the Information Quality Act were all aspects of a deliberate, multi-faceted, and strategic information management approach that nonetheless attempted to be practical and usable for a variety of participants and stakeholders.

1 Introduction

The quality, objectivity and credibility of climate assessment reports are pillars of trust for the development of large-scale policy. The International Panel on Climate Change (IPCC) as well as the first two US National Climate Assessments (NCAs) had significant impact on national

This article is part of a special issue on "The National Climate Assessment: Innovations in Science and Engagement" edited by Katharine Jacobs, Susanne Moser, and James Buizer.

✉ Anne M. Waple
awaple@secondnature.org

1 Second Nature, Boston, MA, USA

2 Cooperative Institute for Climate and Satellites-NC, North Carolina State U/NOAA's National Centers for Environmental Information, Asheville, NC, USA

3 NASA Goddard Space Flight Center, Greenbelt, MD, USA

and international policy-related discussions. These assessments have provided critical 'state of the science' summaries, and the major conclusions of these reports have been well-vetted during the assessment process in addition to resting on peer-reviewed academic literature. The information base for these assessments necessarily emphasizes mature datasets and high confidence information. It is a time-consuming, expert-dependent approach and results in an exceptionally rigorous suite of information.

This paper focuses on the development of an information system for the Third NCA (NCA3) in the context of a broader information system for the US Global Change Research Program (USGCRP). One of the key conclusions of the effort is that success lies not only in the quality of the data and analysis itself, but in a human-centered (not just technical) approach to information management, access, communication, and technical support.

1.1 Background

The first two NCA processes as well as most IPCC assessments can be termed 'summative' assessments, aiming to summarize our current collective understanding by examining both the breadth and depth of existing scientific analysis to take a "snapshot" of what is known at a particular point in time. This 'state of the science' is rigorous and designed to support policy decisions and identify gaps in scientific understanding that can inform research agendas and funding priorities.

Developing a rigorous understanding of what we know is involved and time-consuming. One explicit goal of the NCA3 was to improve the ability of the USGCRP (the interagency program responsible for developing the NCA) to meet the legal requirement to produce a U.S. climate assessment every four years.[1] Meeting this deadline has proved difficult given the enormous effort involved and in order to facilitate timely NCA reporting, a sustained assessment process was developed as part of the NCA3 process (see Buizer et al. 2015). A more rigorous, accessible and well-managed information management approach was needed to support this ongoing effort. In addition to enhancing the efficiency and documentation of the assessment process, this system needed to support the deliberations of the authors as they developed drafts, provide traceability and verification of information sources, and build infrastructure to facilitate current and future assessments.

Additionally, the NCA3 process was designed to serve a wider array of stakeholders. With this in mind, a complementary approach to summative assessment, is a 'formative' style of assessment that allows a more integrated and iterative cycle of information development - from input to usability. Following a distinction used in educational assessment language (e.g. GWAEA 2010), a formative assessment approach allows continual feedback and adjustment, in conjunction with the student, to improve learning outcomes. When applied to scientific assessment, this approach involves stakeholders in the information development process (not just the end-use of information), includes a wider variety of information sources, and an accessible and non-expert interface for contribution and retrieval of the information. While the NCA itself was not designed to provide high-resolution information tailored to all decision-makers, by designing a more inclusive and iterative assessment process, the types of questions being asked, and therefore the types of information brought into the assessment, can be more informative to a wider array of stakeholders. Additionally, the information system used to support such a process, should then

[1] Legal reference: Global Change Research Act of 1990, 15 U.S.C. §§ 2921–2961: http://www.gpo.gov/fdsys/pkg/USCODE-2011-title15/html/USCODE-2011-title15-chap56A.htm

link from a national assessment process to datasets of increasing resolution housed in the agencies and beyond, including in other portals such as the Climate Data Initiative and climate.data.gov.

Therefore the NCA3 had to resolve a tension between serving the information needs of a rigorous and authoritative process designed for policymakers and scientists (and a legal obligation for the NCA to meet the highest tier of the Information Quality Act [IQA]), and providing information that is more timely and relevant for decision makers. The NCA3 approach required a broad re-thinking of the way that information was managed both within the assessment process and "behind" it – a more accessible way to connect to the data that supported the conclusions. NCA3 coordination staff and agency program managers therefore designed several innovative solutions, including the development of a Global Change Information System, a foundation for data management and access.

2 New information management approaches: motivation and framing

2.1 Broad motivation for a federal climate information system

Over the past two decades there has been much discussion about creating a U.S. National Climate Service (NCS). An NCS has never emerged as a formal enterprise,[2] however, these discussions, along with emerging requests from stakeholders, have produced clear calls for access to climate related data and tools through organized federal efforts.

For example, congressional testimony of Deputy Oceanographer of the Navy Robert Winokur, on June 22nd, 2011 indicated:

> "*The Navy desires access to readily available, reliable and consistent data and information in an easily available and preferably consolidated location to move away from the current disparate method of locating and obtaining climate information…*"

And at the National Climate Adaptation Summit, Dr. John Holdren, Director of the White House Office of Science and Technology Policy, issued "seven priorities for near-term action" in 2010, including:

> "*Creating a federal climate information portal. This would provide single-point access to data from all relevant federal agencies and programs and would evolve over time into a more 'national' portal with information about relevant non-federal efforts.*"

Similarly, a recent analysis of stakeholder groups across three western states included a conclusion that "While sometimes conceptualized in different ways, a centralized location or source to obtain climate-related information was a prominent need…" (Dilling and Berggren 2014).

The problem being articulated above is not a lack of information, but rather a lack of accessible and 'organized' or curated information. For example, a coastal manager looking for information on sea level rise and associated impacts may find relevant information in as many as five or more agencies, all of whom have scientifically accurate information that is analyzed and presented differently (using different base periods, data sources, scales, analytical techniques etc), and not necessarily available in formats that are translatable for non-expert users.

[2] In lieu of a consolidated NCS, regional climate services have increased extensively in several USGCRP agencies

In response, and in the context of the development of the NCA3 and its associated data management needs, the USGCRP began to examine the possibilities for meeting the needs of stakeholders and scientists alike who wanted better access to data that was organized and usable.

2.2 NCA as catalyst and pilot

In 2010, stakeholder 'listening sessions' were held (USGCRP 2010a, b) that were designed to contribute to a new strategy for climate assessments. These sessions included a variety of what were considered 'non-traditional' assessment stakeholders (e.g. city sustainability leads, water managers, tribal representatives, private sector attendees) as well as many scientists and federal managers who would ultimately be engaged in supporting the NCA. During these initial workshops, there was considerable focus on how to incorporate non-federal and non-academic data and information into the assessment (e.g. tribal assessments, adaptation data from cities, specific case studies). An unequivocal workshop conclusion was that stakeholders must be part of the process and that there must be a solution that facilitates use of non-academic data and information while still meeting the IQA.

Other listening sessions throughout 2010 echoed similar themes and a specific 'knowledge management' (KM) workshop was held in September 2010 to address the potential solutions to these challenges (USGCRP 2010c). The workshop covered a wide range of topics and included participants from government, NGOs, private sector, and academia – both data experts and expected users. This workshop and its subsequent report were foundational in setting the strategy for information management in the NCA3. The report covered information needs and solutions ranging from how to manage the process of public comment, to considerations of web deployment of the NCA3, to peer review and data archiving. Without this workshop, the NCA3 would have been less well informed, but also its principal actors would have had much more difficulty in effectively communicating the needs to government leads; importantly, the workshop offered a 'line of sight' from the proposed NCA information management strategies to the community input that informed them.

Of note in this KM workshop was a request that the NCA3 'not forget about the authors' in the process of managing information. The contention was that we should not only focus on the needs of the intended users, but also think about how to make the process and the information flow easier for the authors and contributors. This influenced NCA3 priorities on several fronts. For example, this request resulted in the design of a digital author workspace for each chapter team. A balance between employing 'new' collaborative technology and ensuring that it was straightforward to use for all authors (including those unfamiliar with online collaboration) proved absolutely essential and a key to its success. This online author space also served as an integral part of the end-to-end information flow that improved the capturing and tracking of sources and provenance of information and the progress of each team.

Overall, multiple elements of an end-to-end information system were considered and alternatives discussed within the staff, federal agencies, an interagency working group, and NCA advisory committee. Most of the development work was done within the NCA Technical Support Unit at NOAA's National Climatic Data Center. The final elements of the information system included: 1) a portal with data quality and documentation guidance for external contributors, 2) more dynamic access to scenarios of climate and sea level, 3) an author collaboration platform, tracking evolving images and data, 4) a site for managing public comments and author responses, and 5) the NCA3 report itself as an electronic publication, including clickable access to data behind the report.

There were many lessons learned in the process and not all elements were equally successful in execution. However, systematically linking each piece of the information management challenge together and striving to make significant improvements in each piece was a valuable and generally successful effort requiring much new problem solving and innovative solutions.

Given the clear needs and goals of the NCA3, the breadth of data and information that would be incorporated, and the emerging plans for meeting those needs, the USGCRP decided to take advantage of the NCA3 to begin an effort to meet the broader information access challenges articulated in section 2.1. NCA3 became a pilot effort in the development of the Global Change Information System (GCIS), which would use lessons learned in the NCA3 process to build an interagency federal climate information management system. The NCA was ideal for this purpose especially given that it involved a subset of federal climate data, but broad-based enough that it would test out many components of the challenge. Initially, the GCIS was described broadly, encompassing all the components of the end-to-end system as outlined above. Over time, the GCIS came to mean a subset of the full information approach (specifically, the hidden 'back end' of the technical system dealing with traceability and the management of metadata). Lessons learned regarding the definition and communication of the GCIS can be found in section 4.3.

It should be noted that while the GCIS was a new effort and was focused initially on meeting emerging needs, especially associated with the NCA, it nonetheless built off a requirement in the GCRA to periodically recommend guidance for the integrated and accessible management of global change data (section §2934 (d)), and to provide integrated research information to international parties (through a Global Change Research Information Office; section §2953). Over the last two decades, there have been a variety of efforts to coordinate data across the USGCRP agencies in response to these requirements, but for a large number of reasons, it has been a continuing challenge. By focusing on the NCA in its initial phase, the GCIS sought to overcome some of the overwhelming scale issues and develop a sustainable system that could be flexibly built out over time. However, much important experience preceded and informed the GCIS.

In addition, in March of 2014, the Climate Data Initiative (CDI) was launched by the Obama Administration to provide a clearinghouse of climate decision-relevant data, especially to spur private sector innovation around developing climate-related resources from those data. It was followed by the Climate Resilience Toolkit (CRT) launch in November 2014 to provide access to data-driven tools, information and subject-matter expertise around resilience objectives. The CDI, CRT, and NCA are three complementary flagship products providing climate-related information, data and tools. The GCIS, building on the prototype established to support the NCA is being used to integrate and link across those products and later, additional federal data and information.

3 Elements of the NCA-focused information system

3.1 Objectivity, integrity, and utility - the IQA

The Information Quality Act (IQA) (also known as the Data Quality Act) was adopted in 2001 as part of an appropriations bill. It directed the White House Office of Management and Budget (OMB) to guide federal agencies in ensuring that federal data and information is of maximum quality, objectivity, utility and integrity (OMB 2001, 2002; NOAA 2014). According to the criteria for applying the IQA, the NCA falls into a 'highly influential' category and must meet the most stringent IQA requirements.

Meeting IQA guidelines drove much of the conversation around how to guide authors, document processes, and evaluate appropriate use of the data, including non-traditional sources of data and information. The IQA also spurred guidance to authors – most of whom were not federal and not familiar with the IQA – on how to evaluate information on the basis of the four outcomes (quality, objectivity, utility, integrity [USGCRP, 2011]). This was particularly important as they considered information that did not have 'built-in' indicators of quality such as having passed through academic journal peer review.

As soon as the NCA Development and Advisory Committee (NCADAC; the federal advisory committee) was established in February 2011, several ad hoc Working Groups were formed to make some immediate recommendations for NCA processes. For example, Working Group 4 (WG4) was formed to examine "web deployment, peer review, and data management." This was a critical initial working group - (previous federal advisory committees had not so explicitly addressed these issues) – and ensured that the NCADAC itself was an integral part of information management solutions. Initial recommendations from WG4 were informed, in part, by the listening session workshop reports and internal USGCRP discussions, and were submitted to the NCADAC and NCA staff in May 2011.

Forming WG4 was a critical step in the process of identifying priorities for NCA information management that cut across NCA staff, federal agencies, and the federal advisory committee. As a result of WG4's recommendations, a more formal working group of the NCADAC was established in May, 2011 (WG6) for the purposes of evaluating and providing guidance on peer review, information standards and access.

There were several important outputs from WG6 including a Frequently Asked Questions document (FAQs)[3] to help external contributors to the NCA determine how they can best ensure their input would be acceptable for use. Additionally, WG6 examined the likely types of data and information that might be received as part of the external submissions and provided a decision tool that the authors could use in the determination of *how* to use the input appropriately. (See Fig 1). For example, the tool helped determine when specific information could be used as supporting evidence for a key finding versus when it would more appropriately be used only as as an additional illustration of a point, but not as a primary source and not to support a key finding.

These tools and approaches gave the authors the ability to use a wider variety of inputs to the NCA without having to make potentially inconsistent judgment calls as to their value and quality, while still meeting the requirements of the IQA.

3.2 Managing external contributions and comments

As has been mentioned, a process to facilitate a greater degree of practitioner and non-academic information was called for in NCA3. After deliberation within the NCADAC, the USGCRP, and NCA staff, one approach to expanding access to this information for the NCA was a public 'Request for Information' (RFI), issued through a Federal Register Notice (FRN), and further publicized by participants and by the NCA website. This FRN solicited 'technical input' and provided timelines, guidelines (including webinars for those who intended to contribute), and contact information to facilitate follow-up.

This process generated over 200 individual contributions from NGOs, academic groups, federal agencies and others (over 500 documents total). Some of the inputs were extremely comprehensive and analytical involving dozens of team members, and others were brief

[3] http://www.globalchange.gov/sites/globalchange/files/nca-info-quality-assurance-faq.pdf

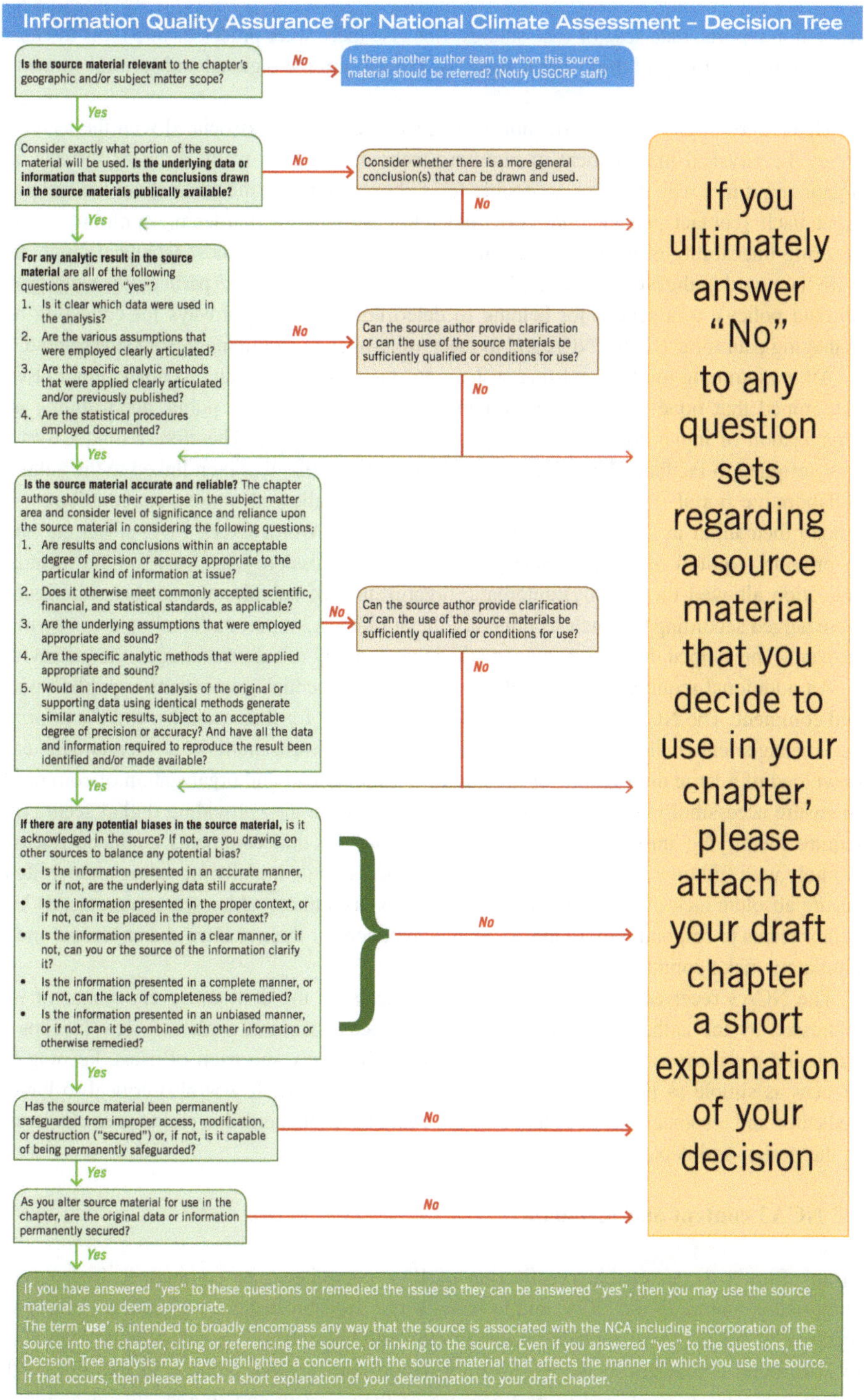

Fig. 1 Decision Tree that was produced by WG6 to guide authors in using external data and information

workshop reports, data sets, or case studies. Because there were processes and tools to ensure appropriate use, many of these inputs were used to support the regional and sectoral content in the NCA3.

There were three major information management challenges associated with the technical input: the physical management of ingesting technical input in many different formats and organizing it in a way that was accessible for authors; gathering metadata about the technical input itself (contact person, title, copyright, etc.) without overburdening contributors; and ensuring traceability of images, statements, data etc. embedded in these technical inputs that were then used in the NCA3. The GCIS Interagency Working Group, (particularly a sub-group on data policy), was central for helping to determine different pathways for collecting and managing data for use in the NCA and to meet multi-agency requirements for stewarding data.

After examining multiple existing systems for managing public and technical input, it was determined that none quite met the criteria for simplicity of use, public accessibility, and organization of inputs for subsequent evaluation by the authors. Therefore a unique system was designed to facilitate the technical input management that was then linked to the authors' collaborative portal. For example, a simple web form allowed non-expert contributors to upload their input in a variety of common formats (MS Word, Excel, PDF, JPEG etc), and to provide required metadata elements about their submission. This very simple metadata collection allowed the (mostly) automatic organization of submissions so that submissions were tagged according to which section(s) of the draft NCA3 report outline they might be most relevant. This helped the authors manage the large quantity of information they had to review.

An additional requirement of the IQA for Highly Influential assessments is public review and comment. The NCA3 was available for public review for three months, and the information management challenges associated with this element were multiple: a) facilitate the easy download of a large draft report for the public and the upload and organization of comments, b) ensure accessibility for all members of the public while also providing digital security, c) require identifying information about the commenter so that comments could be released publicly with full transparency at the end of the process while initially providing anonymity to ensure absolute lack of bias (and perception of bias) as authors addressed comments, d) help author teams to respond only to the comments intended for their chapter and expertise, and e) ultimately make comments and responses public.

The NCA3 received over 4000 separate comments on the draft and every comment was addressed by the authors. The public comment process occurred with very few hiccups or flaws and a critical aspect was prior awareness of the likely spectrum of users, keeping the process as simple as possible while still functioning properly. It was also critical to have a dedicated and responsive team so that changes could be implemented in real time as challenges in the design and deployment arose.

3.3 NCA3 content and metadata

In an effort to better enable certification of IQA compliance, to continue to improve the authoritative nature of the NCA, and to better serve decision-makers, a significant effort was made in NCA3 to provide greater traceability to underlying data. This was possibly one of the most difficult challenges in the NCA3 information management process, in part due to the volume of datasets used, and in part because of their variety and sources. For example, as shown in Fig 2, it is sometimes the case that, even in just one figure, there can be multiple datasets, analyses, scales, and contributors.

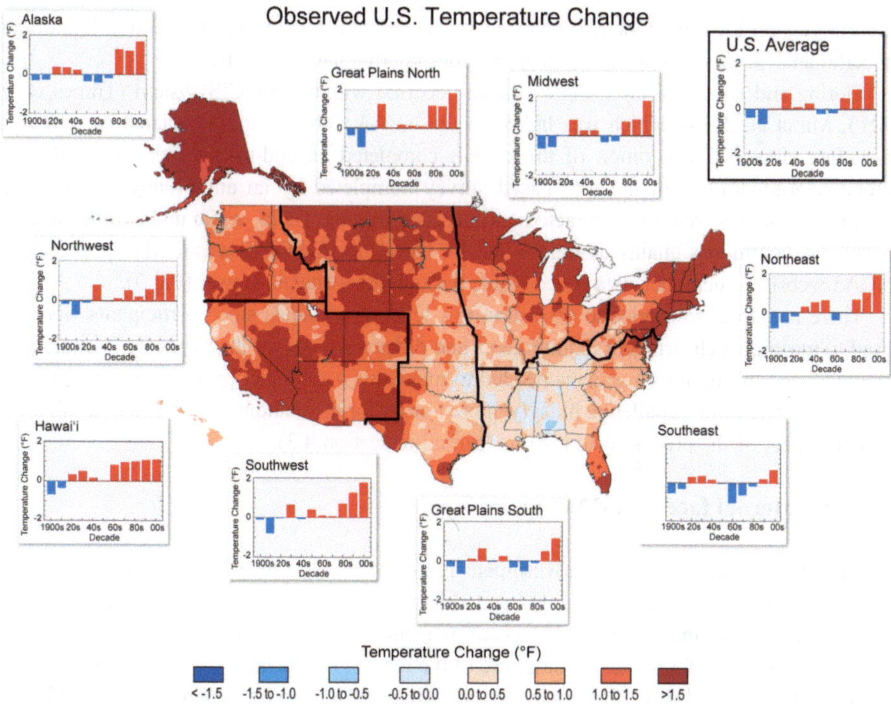

Fig. 2 From Fig. 2.7 in the NCA: 'Observed U.S. Temperature Change', this single figure has 11 images, 2 datasets used in multiple combinations across the 11 images, and nearly 300 metadata field inputs documenting this graphic

Providing traceability from key conclusions to the data that supports those conclusions requires adequate metadata (data about the data). Challenges in building an information system that accommodated this requirement included: (a) the actual development of the metadata (data sources, analysis methods, etc), and involvement of the report authors in providing and reviewing it, (b) documentation demonstrating that the metadata met federal requirements, and (c) development of tools to communicate the metadata to the readers of the report.

The task of managing the metadata included the efforts of a dozen personnel and took over two years from start to finish. It was necessary to translate the metadata into clear and correct documentation, which meant that the information had to be verified. The IQA guidelines require the ISO 19115 metadata documentation standard (NOAA National Coastal Data Development Center 2014). The metadata collection inputs included an effort to answer the "who", "what", "when", "where", and "why" of the NCA3 figures and datasets, and in a manner that also demonstrated transparency and reproducibility.

Initially, basic definitions for metadata inputs, such as "graphic", "figure", and "image", were assigned, and a metadata repository was built within which to store the collected metadata and in the correct format. Using a web-based survey, first tier information about the data was collected (e.g., "where did the data come from?"), by documenting whether a figure, image(s) were original for the report or cited from existing sources and then the specific figure, image and data source information (e.g., named individuals, URLs, publications, etc). In the same web-based survey, other information was collected such as dataset accessibility, dataset version identifiers, named analysis and visualization software and version identifiers, specific analysis and visualization methods, and intermediate products.

Storing and communicating the metadata occurred within the GCIS system (Tilmes et al. 2013, Ma et al. 2014), which was linked to the NCA3 web site and associated custom-built metadata viewer. The addition of these tools completed an end-to-end metadata collection process: originating with the web-based survey completed by an authoritative contributor, completed surveys were then extracted and ingested into the GCIS, while undergoing simultaneous and continuous quality assurance checks. Once met, the GCIS was used to populate the NCA3 website, where metadata is accessible through the metadata viewer (Fig 3).

There is natural resistance to providing this level of metadata since participants were not used to meeting such stringent requirements and did not always see the necessity. Not only are technical and timing improvements needed to make this a smoother process moving forward, but a communications challenge must be met in order for the authors to feel as though this is a valued aspect of the process (see lessons learned in section 4.3).

3.4 The external face of the NCA3

The NCA3 was the first major government report of its kind to be released primarily as a digital report and resource. The ongoing tension between serving policy and science audiences, and serving the public and decision-makers was strongly felt in this dissemination phase. For example, for some, the structure of the report (via the table of contents) historically has been the expected way of accessing the information. In contrast, experience shows that the majority of web-based readers prefer to jump to information by topics that most interest them, and access information in smaller "chunks." In developing the interactive web-deployment for the NCA3, the reader is presented with an interconnected array of information; by clicking on links you can either go deeper into a subset of information (for example to download the data or metadata behind a figure) or away to other related but higher level information. The design in this case created a layered and more dynamic approach to finding and accessing the information while still giving readers the option of downloading a pdf.

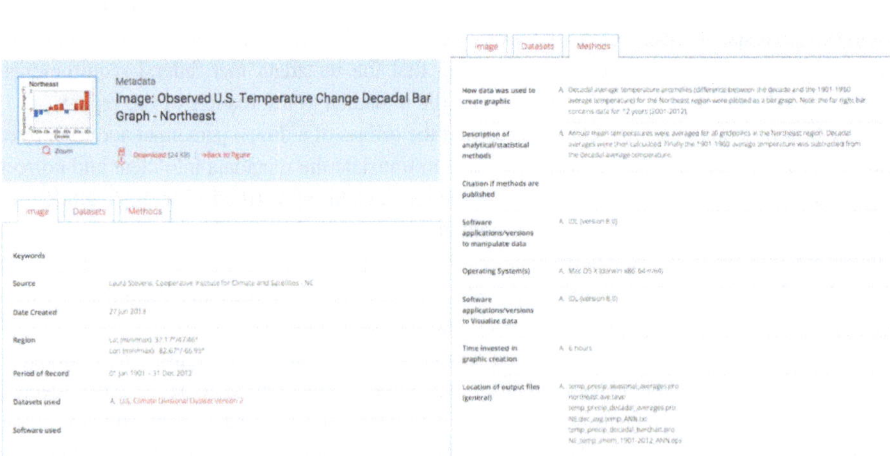

Fig. 3 A sample of the collected metadata for the NCA3 Fig. 2.7, via the NCA3 website metadata viewer. This view is in the "methods" tab of the viewer, where details of data analysis and visualization are provided. (Figure Source: CICS-NC)

Additionally, because the NCA3 information system was built on the premise of an ongoing or "sustained" assessment effort (see Buizer et al. 2015), it is assumed that information will be added to the site before the next periodic synthesis report is developed. Therefore the design has to accommodate the evolution of the information beyond the static format typically found in a downloadable report (developing interactive indicators for example) but also protect the highly vetted NCA3 contents from any subsequent, less-vetted data created for other purposes.

The design of the NCA site (nca2014.globalchange.gov) attempted to balance these priorities and allows users to organize the material by key findings (highest level summary), by region, by sector, by response options, and by a summary level of all sections, or a deeper discussion of the content (the full content of the report). It also allows word searches across the report, and uses the latest design principles to display the information. The highly aesthetic format of the website has been very well received, with positive comments received from a wide range of individuals – from authors, educators, Congressmen, and members of the public to the Chief of Staff in the White House. Readers can jump from highlights to deeper information at any point while browsing the information, and there is seamless connection to the metadata viewer for the rich information behind the figures and graphics.

4 Beyond NCA3: summary of lessons learned from the GCIS

As with any significant emerging project or effort, the specific pathway for design and outcomes are not well known at the outset and designing an information system with such high visibility required a leap of faith. The previous approaches to development and deployment of NCA reports had the advantage of being well described, conceptually clear, and technically proven. Building a new approach to information management increases uncertainty, leading to anxiety among government officials and participants, even though there was strong support for a more innovative, useful, and cost-effective product.

Multiple lessons that are informing the ongoing assessment process resulted from the challenges described above.

4.1 Risk and reward

Many concerns regarding the quality of NCA information sources, the integrity and security of the information systems, and the volume of data and information, stemmed from the NCA3 commitment to an open and inclusive process. As with all inclusive processes, there is risk. Facilitating external contributions was viewed as necessary not only to improve the quality of the assessment, but also to enhance its utility for decision-makers (see Jacobs and Buizer 2015). But it also could have opened the process to an array of negative consequences, including "hacking" from external parties or inadvertent inclusion of information that did not meet IQA standards, among other concerns. Therefore, some of the preparation and implementation for the NCA3 information management system included significant attention to ensuring that the system itself was robust, and that the contributions relied on were of high quality. In reality, the NCA3 process received very little information that was deemed unusable (though not all of it was ultimately used) and there was no evidence of anyone engaging with harmful intent. However, preparation for this possibility was essential to the process (and will be for future assessments). Managing the risks of an open process requires pro-active caution and attention and innovative management from staff and technicians as the systems are assembled. It also requires clear and frequent communication to decision makers in the federal process so that the concerns about and perception of risk did not exceed actual vulnerability (see also section 4.3 below regarding lessons around communication). In reality this can include much more time and energy interpersonally than it does technologically, and budgeting for this aspect of the effort is essential.

4.2 Guidance and training

One of the main lessons from the NCA information management process was that where guidance and training was provided to authors and contributors early and clearly, the process was much more streamlined and ultimately more successful. For example, while guidance for technical input contributors was as comprehensive as it could be at the time, there is still room for improvement in many areas, including in better defining requirements for metadata documentation associated with technical input. There were many other areas where guidance existed but was provided too late to be widely used by contributors, as in the instructions on how to use climate and sea level scenarios (see Kunkel et al. 2015). In still other instances, for example, guidance related to documenting levels of certainty (see Moss 2015) the guidance changed over time (in response to feedback), which authors found confusing.

With the scenarios guidance, timing was the primary issue – the scenarios were not available at the time that the request for technical input was released and so technical input contributors did not have the benefit of consistent future scenarios in their submissions. This also became problematic for integrating information from multiple technical input sources. In terms of the metadata associated with technical input, lessons learned through the process allowed us to better understand how, and importantly *when*, to ask for metadata in future iterations. For example, it is challenging to know early in the process which graphics will be used as support for major conclusions. However, the identification of key graphics (and datasets) and the process of requiring appropriate metadata requires further discussion prior to initiating metadata collection for the next synthesis report.

Unfortunately, most of the metadata for the NCA3 was requested after the report content was mostly assembled. This was at the height of last minute revisions and reviews and the all-volunteer army of authors did not always have the capacity at that time to chase down details of figures and datasets. Two improvements have already been suggested in this regard. The revised process separates the collection of the required metadata information into two tiers. Tier 1 information collects graphic source information (e.g., "Who is responsible for this graphic, and where did it come from?") and is a required first step of report production. Tier 2 information collects reproducibility-level inputs on dataset(s), tools, and analysis methods; this tier is required if there is data and/or analysis included in the visualized results. Metadata collected in the Tier 1 and Tier 2 surveys will be required before a report can proceed to government review. This revised process is already being applied to a special report on climate change and human health that is under development. And to assist in the execution of this new approach, communication is key, especially to those participants who are not familiar with federal data requirements. As part of the revised metadata process, the staff is emphasizing early and often the importance of metadata as not only a Federal requirement, but a "best-practice" throughout the community of climate data production and use.

Where relatively comprehensive guidance was produced (e.g. the IQA source information decision tree), these help clarify processes, risks or questions, and actions. Though these tools took us a long way down a path of facilitating the use of new data and the participation of new stakeholders and still ensuring a robust and rigorous process, it is not clear that many authors internalized this guidance in the array of other guidance that they received. A possible next step with the source information decision tree may be to make it a more interactive digital evaluation tool. This may make the evaluation process of new kinds of information a little smoother and quicker.

Nonetheless, an important outcome of engaging in discussions of using 'new' sources of useful information, and particularly the level of detail with which this discussion was approached, has likely resulted in increased acceptability of using non-government, or non-academic data and information where appropriate. This is a marked shift from previous assessments. The NCADAC working groups played an enormous role in ensuring that this discussion was incorporated routinely into deliberations and approached systematically. The conclusions and guidance that were then generated 'internally' to the NCADAC became more acceptable and influential as a result.

4.3 Communication and scale

Delivering the NCA3 electronically was a major decision for the NCADAC and for the federal agencies, and it is likely to have lasting repercussions. This decision was made at the highest policy levels, following the broad expansion of information technology options and a commitment of the Obama administration to innovation in this area. Still, the volume of material involved and the challenges discussed in this paper made this a daunting effort. In the end, this would not have been accomplished without the support of external contractors in addition to extremely dedicated staff in both the USGCRP coordination office and the NOAA Technical Support Unit.

A key to successfully moving into electronic delivery was constant communication and clear articulation of benefits and risk. There remained a high demand for detailed information regarding the specifics of the NCA3 delivery throughout its development and although it was a time-consuming process to keep multiple groups apprised of possibilities, options, or decisions regarding information deployment, it was an essential element, particularly as there was no precedent.

However, the success in linking the online delivery of the NCA3 and the data and information support system behind it has not yet been fully realized (though is continually improving). There was some concern about the GCIS for a variety of reasons, resulting in two important lessons:

1) The GCIS initially was described to encompass a full system – from 'front end' web deployment of content and access, to the deep 'back end' servers and systems to manage and connect all the information. This full GCIS system was difficult to describe succinctly, and the scale was concerning to some officials whose support was critical to progress, but whose deep understanding of the system was inevitably limited. Over time, the specific description of the GCIS became constrained to the functions managing the traceability and metadata – the 'hidden' technical component. However, while it was easier to communicate to non-experts about the system by focusing on only one or two specific components at any given time, the vision of a truly end-to-end approach that connects people and input all the way through to deployment, managing multiple strands of content, metadata, graphics, and more along the way, was still essential in developing a coherent *design*. This tension may always be present to some degree, and paying more attention to smart communication may be necessary as the program moves forward, and as it requires further decision-making.

2) The second challenge of the GCIS was that it was perceived as a 'challenge' to some individual agency-managed web or data responsibilities or opportunities. Although there was explicit communication that GCIS would not house any actual data (this would still

happen in the agencies), and it would not seek to displace users of individual agency websites, building credibility and investment in a shared system was difficult, especially in an era of limited resources. More could have been done initially to bring agencies together to build common goals and shared commitment, and this will need to be an ongoing element of the GCIS build out, even as its value is increasingly recognized. An interagency GCIS working group was an important part of communication across the agencies, and further work to ensure this group can help to communicate at appropriate levels in their agencies would be very valuable.

4.4 Capacity and ownership

While this article has not discussed staffing in any detail, the bulk of the work was accomplished by the NCA Technical Support Unit (TSU), and by the NCA staff at the USGCRP, with important input from a multi-agency GCIS Interagency Working Group. These teams were heavily interconnected and much of the early development of the NCA information management strategy required daily interaction and strong core relationships between the teams.

It is essential for the success of USGCRP and NCA information management that there is collective investment (financially and professionally) in the process, and the opportunity to build long-term institutional memory and clear responsibility and accountability. While NOAA requested and received specific funding for the TSU, there must be an ongoing understanding of all of the agencies' roles and a clear path of communication and responsibility to the NCA coordination office and NCA Director in the future in order to build from the current foundation to a fully functional information system and to more easily deploy future assessments.

5 Conclusion

The information management challenge for the NCA3 was significant. Not only did the process seek to improve the ability to meet the existing imperatives for the assessment, but also intended to serve new stakeholders and become a sustained (and sustainable) process. The electronic delivery of this report, along with data transparency and traceability, though in retrospect among the most dramatic successes, was also a major cause of stress throughout the process.

It was essential to success that the goals of the NCA3 included the clear intent to become a long-term sustained process. Without this understanding and motivation, the scale of solutions would have been smaller, less mindful of multiple ongoing circumstances and users, and less designed for stability and re-use. The products and services for NCA3 information management were more robust as a result of understanding them to be part of a long term process and larger scale solutions.

It was also important to consider the solutions as all being part of a coherent, linked end-to-end system beginning with user requirements as the driver. How *people* need to use the system was the critical first step and overarching consideration. The technological solution then could focus on the balance of maximizing simplicity with credibility, cost and utility considerations.

There are important lessons learned and continuing improvements that are already being made, including building on the successes with better training and guidance for participants, better timed collection of data, more interactive usability for tools that were valuable, and frequent and appropriate communication to a stakeholders and decision makers. However, an important conclusion of this process is that a tremendous amount of progress can be made in a relatively short period of time if there is stakeholder input from the beginning, there is an inclusive process for moving forward, a dedicated team of problem solvers that has support from leadership, and a laser-focus on the three primary end users of the system: the American public, on-the-ground managers and their elected representatives.

Acknowledgments This paper is an assessment of the information management process associated with the NCA, however the Global Change Information System project included many key contributors and a wide variety of input. While it is not possible to name all of the GCIS contributors, those who played critical roles include: Anna Pinheiro Privette, April Sides, Steve Aulenbach, Andrew Buddenberg, Glynis Lough, Robert Wolfe, Brian Duggan, Justin Goldstein, Angel Li, RPI Tetherless World Constellation staff, NEMAC, Habitat Seven, the GCIS Interagency Working Group, the NCADAC, and the staff of both the NOAA Technical Support Unit and the USGCRP Coordination Office.

This work was partially supported by NOAA through the Cooperative Institute for Climate and Satellites — North Carolina under Cooperative Agreement NA14NES432003.

References

Buizer JL, Dow K, Black ME, Jacobs KL, Waple A, Moss RH, Moser S, Gustafson DI, Richmond TC, Hays SL, Field CB (2015) Building a sustained climate assessment process. Clim Chan. doi:10.1007/s10584-015-1501-4

Dilling L, Berggren J (2014) What do stakeholders need to manage for climate change and variability? A document-based analysis from three mountain states in the western USA. Reg Environ Chang. doi:10.1007/s10113-014-0668-y

Grant Wood Area Education Agency (2010). Literature Review: Assessment for Learning (Formative Assessment). Grant Wood Area Education Agency. http://www.gwaea.org/iowacorecurriculum/docs/AssessmentForLearning_LitReviewFinal.pdf

Jacobs KL, Buizer JL (2015) Building community, credibility and knowledge: the third U.S. National Climate Assessment. Clim Chang. doi:10.1007/s10584-015-1445-8

Kunkel KE, Moss R, Parris A (2015) Innovations in science and scenarios for assessment. Clim Chang. doi:10.1007/s10584-015-1494-z

National Oceanic and Atmospheric Administration, (2014), revised. National Oceanic and Atmospheric Administration Information Quality Guidelines [Available online at http://www.cio.noaa.gov/services_programs/info_quality.html

NOAA National Coastal Data Development Center, cited (2014). Metadata Standards. [Available online at http://www.ncddc.noaa.gov/metadata-standards/]

Ma X, Fox P, Tilmes C, Jacobs K, Waple AM (2014) Capturing provenance of global change information. Nat Clim Chang 4:409–412

Moss RH (2015) Assessing decision support systems and levels of confidence to narrow the climate information "usability gap". Clim Chang. doi:10.1007/s10584-015-1549-1

Office of Management and Budget, (2001). Guidelines for Ensuring and Maximizing the Quality, Objectivity, Utility, and Integrity of Information Disseminated by Federal Agencies," 66 Federal Register 189 (28 September 2001), pp. 49718–49725 http://www.uspto.gov/products/catalog/infoqualityguide.jsp

Office of Management and Budget, (2002). Information Quality Act Final Guidelines http://www.whitehouse.gov/sites/default/files/omb/fedreg/reproducible2.pdf

Tilmes, C., Fox, P., Ma, X., McGuinness, D., Pinheiro-Privette, A., Smith, A., Waple, A.M., Zednik, S., and Zheng, J., (2013). Provenance Representation for the National Climate Assessment in the Global Change Information System. Trans on Geosci and Remote Sens. Volume: 51, Issue: 11 doi:10.1109/TGRS.2013.2262179, pp 5160–5168

USGCRP, (2010a). The United States national climate assessment NCA report series, vol. 1: Midwest Regional Workshop: Feb. 22-24, 2010, Chicago, Illinois. USGCRP

USGCRP, (2010b). The United States national climate assessment NCA report series, vol. 2: Strategic Planning Workshop. USGCRP

USGCRP, (2010c). The United States national climate assessment NCA report series, vol. 3: Knowledge Management Workshop. USGCRP

USGCRP, (2011). Inf Qual Princ, at http://www.globalchange.gov/sites/globalchange/files/Information-Quality-Principles-Draft_2011-11-16.pdf

Climatic Change (2016) 135:85–96
DOI 10.1007/s10584-016-1609-1

Building an integrated U.S. National Climate Indicators System

Melissa A. Kenney[1] **· Anthony C. Janetos**[2] **·**
Glynis C. Lough[3]

Received: 19 August 2015 / Accepted: 14 January 2016 / Published online: 28 January 2016

Abstract During the development of the Third U.S. National Climate Assessment, an indicators system was recommended as a foundational product to support a sustained assessment process (Buizer et al. 2013). The development of this system, which we call the National Climate Indicators System (NCIS), has been an important early product of a sustained assessment process. In this paper, we describe the scoping and development of recommendations and prototypes for the NCIS, with the expectation that the process and lessons learned will be useful to others developing suites of indicators. Key factors of initial success are detailed, as well as a robust vision and decision criteria for future development; we also provide suggestions for voluntary support of the broader scientific community, and for funding priorities, including a research team to coordinate and prototype the indicators, system, and process. Moving forward, sufficient coordination and scientific expertise to implement and

This article is part of a special issue on "The National Climate Assessment: Innovations in Science and Engagement" edited by Katharine Jacobs, Susanne Moser, and James Buizer.

Electronic supplementary material The online version of this article (doi:10.1007/s10584-016-1609-1) contains supplementary material, which is available to authorized users.

✉ Melissa A. Kenney
kenney@umd.edu

Anthony C. Janetos
ajanetos@bu.edu

Glynis C. Lough
glough@usgcrp.gov

1 Earth System Science Interdisciplinary Center / Cooperative Institute for Climate and Satellites-Maryland, University of Maryland, 5825 University Research Court, Suite 4001, College Park, MD 20740-3823, USA

2 The Frederick S. Pardee Center for the Study of the Longer-Range Future, Boston University, 67 Bay State Road, Boston, MA 02215, USA

3 University Corporation for Atmospheric Research, 1800 G Street, NW, Suite 9100, Washington, DC 20006, USA

maintain the NCIS, as well as creation of a structure for scientific input from the broader community, will be crucial to its success.

1 Introduction

The Third U.S. National Climate Assessment (NCA3; NCA when referring to the effort more broadly) developed new modes of conceptualizing its engagement processes and of integrating evolving scientific insights beyond the quadrennial assessment reports (Moser et al. 2015, this issue). Thus, over the past several years there have been calls for developing a climate indicators systems as part of the NCA (Buizer et al. 2013; Janetos et al. 2012). A National Research Council report (2009a) specifically suggested the U.S. Global Change Research Program (USGCRP) should consider developing a report card or climate change indices that could communicate complex scientific information effectively to a variety of audiences. Such indicators were expected to have utility beyond the NCA; for example, Victor and Kennel (2014) called for a broader range of climate indicators and the National Research Council (2013, 2015) recommended the next strategies for national sustainability to include explicit recognition of decision-making and adaptation and the use of metrics and indicators to evaluate progress.

Indicators are reference tools that can be used to regularly update status, rates of change, or trends of a phenomenon using measured data, modeled data, or an index to assess or advance scientific understanding, to communicate, to inform decision-making, or to denote progress in achieving management objectives. Indicators differ from data visualization tools in that they are systematically updated *and* comparative to a baseline of change.

Indicators have long been used to describe status and assess progress for key phenomena that are measureable directly or indirectly (e.g. through the use of proxies). Some economic and health indicators have been developed and used consistently and effectively across locations, and refined strategically over multiple decades (Cobb and Rixford 1998; Lippman 2006), but not all efforts have been successful (Innes 1990) or institutionalized (Ogburn 1933). Drawing on those experiences, more recent efforts have called for market and non-market valuation of environmental assets (National Research Council 1999) and the development of national ecological indicators (National Research Council 2000) that led to *The State of the Nations Ecosystems* effort developed by the Heinz Center (2008). And recently the National Research Council initiated an effort to consider sustainability metrics (2015). Though physical climate indicators are already presented through a range of Federal agency websites (e.g., https://www.climate.gov/maps-data), a unified effort that could support a sustained NCA process did not previously exist.

In considering establishment of an indicators system during the development of the NCA3, the USGCRP hosted three indicators workshops focused on ecological indicators (USGCRP 2010), physical indicators (USGCRP 2011b), and societal indicators (USGCRP 2011a) (Fig. 1). All the workshop outcomes supported the idea of an indicator system and noted the need for 1) developing processes for a sustained indicators effort, 2) leveraging existing data and indicator products, and 3) including stakeholder perspectives to assure utility for user communities (Online Resource 1: Outcomes of NCA Workshops focused on Indicators).

The workshops started a multi-year process (Fig. 1) to develop recommendations and create a prototype of an interagency indicators system, which we call the National Climate Indicator System (NCIS). The National Climate Assessment and Development Advisory Committee

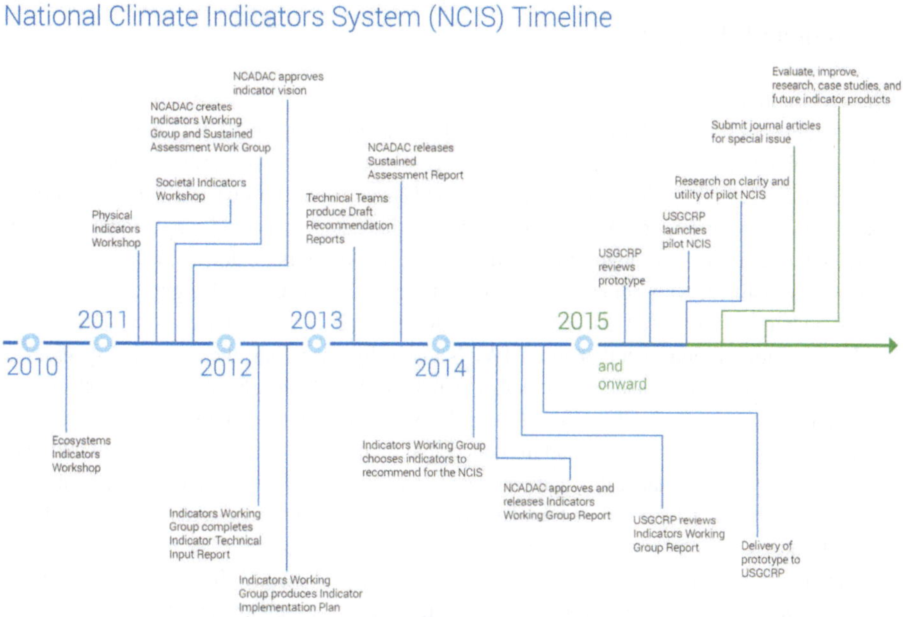

Fig. 1 Timeline for the creation of a National Climate Indicators System

(NCADAC), the Federal Advisory Committee established for NCA3, empaneled an Indicators Working Group (hereafter IWG) in 2011 (Fig. 1; USGCRP 2015), and asked them to develop recommendations (Janetos et al. 2012; Buizer et al. 2013). The IWG included about 25 members from the NCADAC, Federal agencies, academia, and the private sector. It then created 13 Technical Teams of producers and users of information to provide recommendations on specific topics. In addition, an Indicators Research Team, which included both funded and volunteer scientists, coordinated the IWG and Technical Teams (see section 3.1) and developed prototypes in support of the recommendations.

The aspirational goal for the NCIS was that the recommendations of the IWG would be implemented, expanded, and sustained over the long-term to support the NCA and information needs of its users. Additionally, the NCIS was envisioned to be synergistic with Federal climate priorities, including the USGCRP strategic plan and priorities (USGCRP 2012) and later the President's Climate Action Plan (2013); Climate Resilience Toolkit (https://toolkit.climate.gov/); and Executive Orders 13653 (November 2013), 13514 (October 2009), and 13642 (May 2013).

The objective of this paper is to describe the process of visioning, prototyping, and implementing the recommendations for an NCIS and to present lessons learned. Given the recent release of pilot indicators by the USGCRP (www.globalchange.gov/explore/indicators), we discuss opportunities for further development of the NCIS and the research needed for attaining the IWG's vision. The authors of this paper do not claim to be impartial; Janetos and Kenney co-led the NCADAC IWG, Kenney led the Indicators Research Team, and Lough facilitated the review of NCIS recommendations in her role at the USGCRP National Coordination Office. In this article, the term "we" is used to reflect the perspective of the authors of this paper; in all other instances we use the name of the scientific or agency group to minimize confusion about different actors' roles.

2 Vision and decision criteria

The IWG initially produced NCIS vision and decision criteria (USGCRP 2015), in which the vision outlined a system of physical, natural, and societal indicators that communicate and inform decisions about key aspects of the physical climate, climate impacts, vulnerabilities, and preparedness (Fig. 2). Central to this vision was a set of indicators useful for ongoing tracking of critical changes and impacts important to the NCA as well as indicators that could support adaptation and mitigation decisions by a range of stakeholders. These indicators could be tracked as a part of ongoing assessment, with adjustments made as necessary to adapt to changing conditions and understanding (for expanded discussion see Janetos et al. (2012) and Kenney et al. (2014)).

The decision criteria operationalized IWG's vision by describing the types and qualities of indicators that could be considered for inclusion in the system (Online Resource 2: Recommended Decision Criteria for the NCIS; USGCRP 2015). Key criteria include indicators that: are scientifically defensible and useful, support the conceptual framework (Fig. 2), are nationally important (not necessarily of national scale), and encompass lagging, coincident, and leading indicators.

These criteria provided robust foundations for the subsequent recommendation and development phases. Careful evaluation and consensus about the overall objectives, purpose, and criteria for inclusion are essential in an indicator effort to ensure that any competing and incompatible goals are worked through before developing and choosing indicators.

3 Recommendations and development processes

Moving from the overall goals to building the system, the IWG developed a process for making recommendations and creating prototypes that included establishing Technical Teams,

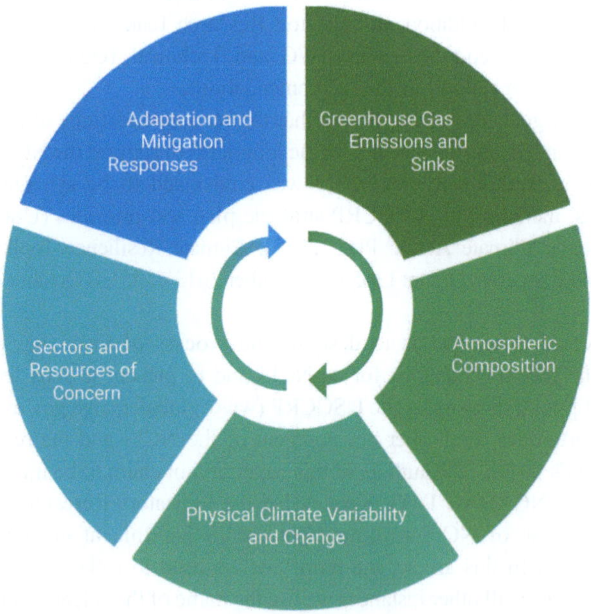

Fig. 2 Conceptual model recommended for the National Climate Indicators System (adapted from Kenney et al. 2014)

creating conceptual models, developing recommendations and research priorities, reviewing the recommendations and selections, and selecting indicators for the NCIS pilot (Fig. 3).

3.1 Creation of technical teams and selection of experts

To develop the NCIS, the IWG created Technical Teams to provide topical recommendations, loosely structured on the sectors and systems represented in the NCA3 chapters (see Table 1). The 13 Technical Teams were comprised of physical, natural, and social scientists and practitioners who had a broad understanding of the multi-stressor climate changes or impacts on the system/sector, experience with existing sustained datasets and indicators, and an understanding of the information needs of stakeholders. In total, the members of the 13 Technical Teams and IWG included over 200 scientists and practitioners who were employed in the Federal agencies (nine of the 13 USGCRP agencies), academia, non-governmental organizations, and the private sector. This range of expertise provided the perspectives needed to meet NCIS goals, to have broad buy-in from the scientific community

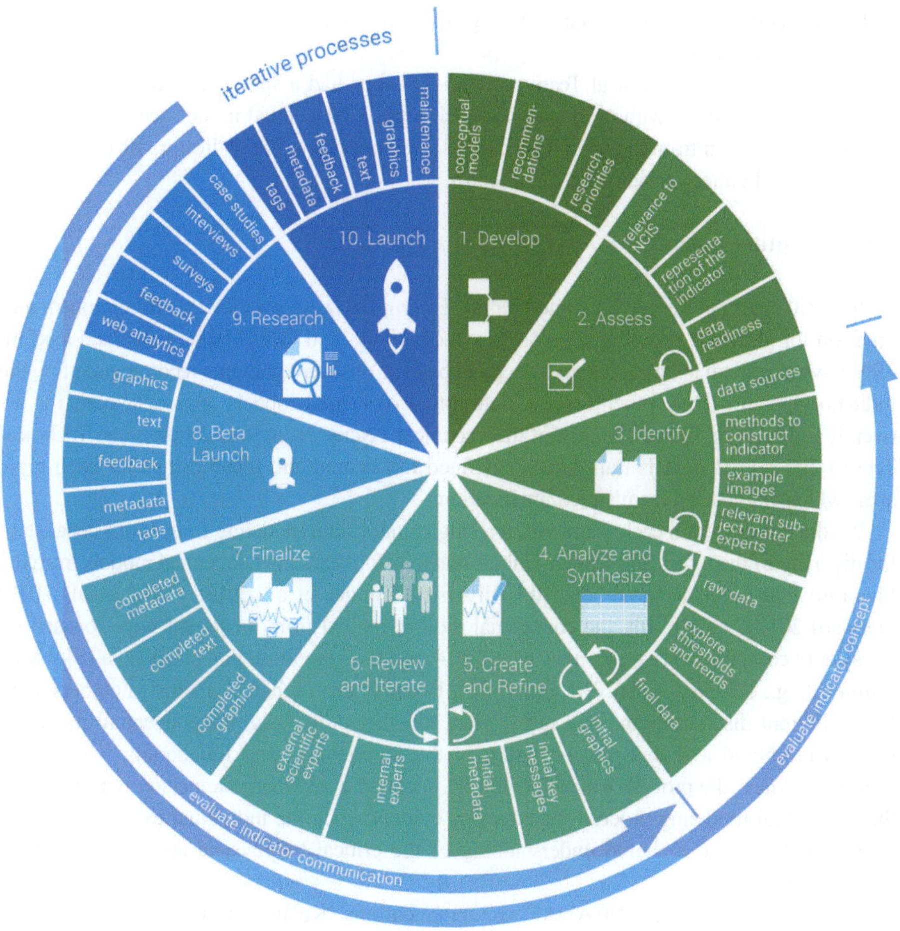

Fig. 3 Proposed and prototyped National Climate Indicators System development and implementation process

Table 1 Technical teams established by the NCADAC indicators working group to develop recommendations for the National Climate Indicators System

Indicators Technical Teams
Physical Climate and Atmospheric Composition
Freshwater Ecosystems
Oceans and Coasts
Water Cycle and Management
Phenology
Forests
Grasslands, Rangelands, and Pastures
Agriculture
Energy
Infrastructure
Human Health
Mitigation and Greenhouse Gases and Sinks
Adaptation and Hazards

and the Federal agencies, and to support the goals of broad engagement within the NCA (Cloyd et al. 2015 this issue).

To ensure that the Technical Team recommendations had a strong theoretical foundation, the IWG charged them with producing for each topic a conceptual model, recommendations for indicators drawn from those models, and research priorities. The Indicators Research Team served a coordinating capacity for these teams.

3.2 Conceptual models

In the 2000 National Research Council report, *Ecological Indicators for the Nation*, one major criterion for indicator selection suggested by the authors (and identified by Cobb and Rixford 1998), was for them to have a clear conceptual basis. Consistent with this recommendation, the indicators system development used a conceptual or mental models approach, both for the overall system (Fig. 2) and for each of the Technical Teams' recommendations. The Technical Team's conceptual models transparently defined the system and relationships for each topic area, which framed the team's recommendations.

We did not constrain each team within a single conceptual modeling framework because of the diversity of the topics and team members and a lack of existing conceptual models from the literature that could accommodate this diversity (Binder et al. 2013; Fisher et al. 2013; Dorward 2014). As a result, the conceptual models ranged from those that were focused on the system components (e.g., Kenney et al. 2014, pp. B-23) to those that provided a risk-based framing (e.g., Kenney et al. 2014, pp. B-42). Because the Technical Teams' members were from different disciplines and had not worked together previously, the conceptual modeling process promoted a shared understanding of the components and functioning within each sector. We knew the process was achieving its intended goal when one person remarked that the conceptual modeling process' success was evident when the individual experts became a team and developed a shared understanding of the critical biophysical and societal factors affecting the sector.

After the release of the NCADAC recommendations (Kenney et al. 2014) and building from the first round of conceptual models, the second round of conceptual model development has focused on using a common approach and syntax. Synthesizing across all the conceptual models has allowed us to better understand and describe the interlinkages between the different systems and sectors, to further describe the overall NCIS conceptual model, and was used to support the web-design of the NCIS. Though the system components are fairly obvious to the Technical Team, they are more challenging to identify and to connect for non-technical experts. We have found the conceptual models to be a useful way to represent systems and describe the types of indicators that could be useful to decision makers.

3.3 Recommendations and research priorities

The large number of existing datasets, indicator products, and graphics provided a solid foundation for the development of the NCIS, including materials created through the development of the NCA3, various Federal agency climate programs, and extramural scientific research. Each Technical Team assessed the available information and made recommendations for indicators that were a balance between those useful for documenting change in the sector and those that would be directly useful for adaptation or mitigation decisions. In either case, the teams needed to demonstrate actual or potential use of the indicator by user audiences to justify their recommendation. At the same time, the teams also suggested research priorities that would build out the information important for each sector.

3.4 Review processes

Each Technical Team's conceptual model, recommendations, and research priorities were developed into a draft report that was reviewed by a minimum of three experts who were not part of the indicators process. The Indicators Research Team managed this external review process, and suggested priorities for responses by the Technical Teams, analogous to standard review processes used in most scientific publishing. The goal was to ensure that recommendations that ultimately came forward to the USGCRP had passed through rigorous internal and external expert review. The revised reports were delivered to the IWG for discussion and synthesis to suggest refinements of the recommendations by a particular team or to identify linkages of indicator and system concepts between teams. In this way, recommendations from Technical Teams underwent three levels of review and discussion – an external review, a review internal to the teams, and a discussion among all the Technical Teams for any system-wide issues, such as overlaps in proposed indicators or unexpected gaps in the end-to-end nature of the system.

Such extensive review was time-consuming and difficult. On one hand, it certainly had the potential to slow down recommendations from coming forward to the NCADAC (and ultimately the USGCRP). However, it created a level of confidence that indicators that were ultimately recommended through this process were scientifically well founded, had a strong potential to be immediately usable, and could be implemented fairly easily.

3.5 Selection of indicators for system

The IWG considered the recommendations of each of the Technical Teams. Because the first phase of the proposed NCIS implementation process was the creation of a pilot indicator system, the IWG asked each Technical Team to provide no more than four indicators that were already scientifically vetted and would be implementable with no additional research or data synthesis. The IWG then developed synthesis recommendations (Kenney et al. 2014), which were approved by the NCADAC (USGCRP 2015), and delivered as advice to the USGCRP. The IWG was well aware that the initial proposals of such a small number of indicators would leave unwelcome gaps in the end-to-end model of the NCIS. However, we felt strongly that it would not be possible to anticipate all the ways in which the indicators of the proposed NCIS might be interpreted and used, and that it would be preferable to make initial progress that could then be evaluated and built on. Technical Teams that had proposed multiple indicators for their topics were thus faced with a difficult tradeoff at the system level of the NCIS.

4 Proposed implementation and phasing

Though the vision, recommendations, and proposed implementation were designed through the IWG, the intent was for the NCIS to transition to the USGCRP for implementation and ongoing management. Specifically, the IWG envisioned that the NCIS would be developed with a process similar to a typical software development cycle. A beta version would be released to the public for review. This pilot version would serve as a proof-of-concept, with a limited number of indicators covering a wide range of topics and providing an opportunity to engage the broader stakeholder community. Such an iterative user-focused approach is rarely employed for decision-support products (Moss et al. 2014), but provides a means to assess the efficacy of the product and to co-design the elements with users to ensure the product better meets their needs (Fig. 3).

To support the implementation recommendations, the Indicators Research Team developed prototypes for a subset of the pilot indicators that were consistent with recommendations from the NCADAC (Kenney et al. 2014). The Indicators Research team also created a prototype style guide for indicator graphics, text, and metadata (see Online Resource 3: Recommended NCIS Graphic Style Guidance).

It is unusual for USGCRP to receive recommended content from an external group. In order to release the pilot as an interagency product, the agencies needed to establish a process to review, revise, and finalize the indicators figures, text, and metadata. A team of Federal scientists reviewed the content for scientific validity, consistency with other USGCRP products (e.g., NCA3), adherence to the USGCRP's communication principles, and correctness and completeness of metadata. A number of changes were made to meet the standards for USGCRP products. For example, trends were added to some figures, text was revised to clarify climate impacts, relevance to decision-making was framed more consistently, and metadata was revised to adhere to agency standards. Several prototyped indicators were removed from the pilot set due to competing agency science products (e.g., multiple methods for calculating drought indices), or because the index reflected non-climate variables and a climate signal could not be clearly distinguished. Due to time constraints, these issues could not be resolved for the pilot, but those indicators could be revisited for the future. Consistent with the NCADAC's recommendations, the USGCRP released a pilot phase in May 2015 (www.globalchange.gov/explore/indicators) and will consider establishing a sustained NCIS that expands the pilot.

5 Co-creation with stakeholders using social science research approaches

One of the tenets of stakeholder engagement processes is that the co-creation of knowledge confers greater legitimacy and salience to the outcomes than simply communicating expert judgment (Moss et al. 2014; Lee 1993; Innes 1990). For the NCIS, the approach to co-creation was serial. The first phase in the NCIS co-production process could be classified as integrating expert knowledge (Meadow et al. 2015). The initial process included experts on the IWG and Technical Teams who were familiar with scientific outputs and broad decision needs; at this stage it did not include non-technical experts who were interested in specific uses for the indicators.

Part of the rationale for the recommended pilot phase was to "road test" the NCIS pilot to understand how the indicators are used and the additional features needed to increase its utility. Involving a diverse group of non-technical experts and stakeholders would create the opportunity to move into a collaborative co-production function (Meadow et al. 2015) to expand and improve the NCIS to meet user needs and preferences.

The two-phase co-development process was purposeful; we focused on planning an inclusive, pragmatic process that would be led by technical experts but informed by decision needs and indicator use. The later phase of the co-development process allows for action-oriented engagement of stakeholders in which people are asked to engage in the process of designing scientific products to increase utility for both the scientific and user communities.

After the pilot NCIS was released by USGCRP, new opportunities emerged to evaluate co-development pathways, including both "top-down" and "bottom-up" approaches. Top-down approaches include web analytics and surveys to provide a broad assessment of the understandability of individual indicators as well as experiments to test the efficacy of visual indicator representations and the system as a whole. Bottom-up case studies engage scientists and decision makers to understand their decision-making contexts, processes, and information needs – particularly, how NCIS indicators or proposed indicators could be modified to better support a range of research questions and climate-informed decisions. The Indicators Research Team is currently conducting and publishing both top-down and bottom-up research that could provide ideas for future development of the NCIS. Such evaluation data can help prioritize indicator modifications, additions, and features for the overall system. An iterative approach to development supports the recommendations by Weaver et al. (2014) to increase global change social science research and National Research Council recommendation (2009b) that future research should include "design and application of decision support process and assessment of decision support experiments."

Through a combination of evaluation research and targeted co-production, the NCIS could be refined and developed in a way that balances the need for indicators to support sustained assessments and to support decisions. Indicators can be both descriptive - describing existing conditions - and analytic - seeking to understand why those conditions exist (Cobb and Rixford 1998). The IWG recommendations primarily focused on identifying descriptive indicators and understanding the relationships among them through conceptual models. The approach placed the descriptive indicators within a systems context without being prescriptive; thus, allowing for a balance between indicators for assessment and decision-making.

6 Research opportunities

To meet the goals of the NCIS vision, several near-term research foci were suggested by Janetos and Kenney (2015). First, a big gap noted in the NCADAC's recommendations (Kenney et al. 2014) was the need for adaptation and mitigation indicators. Because adaptation and mitigation indicators are both more experimental and more closely related to policy than others, the Technical Teams chose to identify research questions rather than individual indicators. For example, further research could help develop indicators that demonstrate climate vulnerability with and without adaptation measures (i.e., counterfactual analysis) or the effectiveness of recently implemented policies to reduce greenhouse gases. A particularly useful set of indicators would move beyond accounting for adaptation actions into measuring outcomes and effectiveness of portfolios of climate-resilience actions.

Second, it will be important to develop strategies for creating leading indicators to support climate-resilient decisions. Leading indicators are predictive of future conditions, based either on currently measured indicators that signify something important about the future or on models that can verifiably forecast future conditions. Leading indicators have been created for a number of key climate variables, such as surface temperature and precipitation (Walsh et al. 2014), but leading indicators for climate impacts are less developed yet critically important for informing adaptation actions. To pursue a research strategy related to leading climate impact and vulnerability indicators, it will be advisable when scientifically feasible to identify their optimal temporal scale so that they can be useful in both current and future decision contexts.

Third, there is a need for spatial scalability, i.e., indicators that can be presented at local to national scales. Such scalability of indicators would allow information to match the scale of the decision at hand. The spatial scalability of indicators is not simply a matter of aggregating or disaggregating observations or modeled results. The most appropriate way of representing an indicator such as drought (Steinemann and Cavalcanti 2006), for example, may depend on the scale, decision context, or available datasets. Overall, a system of scalable indicators would have greater utility for a broader range of users and, beyond the communication potential, could truly support decisions.

7 Conclusions

The recent release of the USGCRP indicators pilot (2015) has been very well received, but the true success of the effort will be demonstrated through ongoing governance and development. In particular, a clear vision of a fuller implementation of the NCIS needs to be defined and refined by user needs and scientific feasibility. In the long-term, a periodic review of the vision and decision criteria could be used to assist USGCRP in meeting, improving, and evolving the goals of the effort. Although the NCIS started with IWG recommendations, the system was intended for Federal agency ownership. Thus, future reconsideration of the goals will likely be a Federal activity, but we believe input from the broader community would prove useful to assure that the NCIS is serving the sustained assessment and user needs.

Reliance on a collaborative and transparent process involving Federal and non-Federal experts and decision makers could guide the evolution of a system of credible, salient, and useful indicators of change that meet the needs of the indicator audiences. In particular, ongoing interaction with stakeholders will allow agencies to better match user needs for climate information with the development of scientifically rigorous indicators. This approach would provide a consistent and regularly updated suite of national indicators to improve understanding of changes and impacts and to strengthen the ability of U.S. communities and the economy to prepare and respond; thus, meeting the original vision of the NCIS.

Ultimately, the NCIS could support development of assessment reports by providing consistent and routinely updated indicators of physical, ecological, and societal change. The NCIS could enable an efficient way for each successive national assessment to build on a well-documented time series of indicators, measured from established baselines, clearly showing how conditions have changed since the previous assessment. Finally, an NCIS that successfully includes information on the climate system, climate impacts, and adaptation and mitigation responses, should be able to provide future assessments with the ability assess the Nation's progress in responding to climate change.

Acknowledgments The authors thank the editors of the special issue, four anonymous reviewers, and Michael Gerst for comments that have improved the manuscript. Kenney's work was supported by National Oceanic and Atmospheric Administration grant NA09NES4400006 and NA14NES4320003 (Cooperative Climate and Satellites-CICS) at the University of Maryland/ESSIC. The National Environmental Modeling and Analysis Center at the University of North Carolina-Asheville, and specifically Caroline Dougherty, Karin Rogers, Ian Johnson, and Jim Fox collaborated with Kenney and Ainsley Lloyd and co-authored the NCIS Style Guide (Online Resource 3). Current and previous Indicators Research Team members include Ainsley Lloyd, Michael Gerst, Rebecca Aicher, Felix Wolfinger, Omar Malik, Sarah Anderson, Julian Reyes, Allison Bredder, Amanda Lamoureux, Maria Sharova, Eric Golman, Ella Clarke, Ryan Clark, Christian McGillen, Justin Shaifer, Olivia Poon, Jeremy Ardanuy, Ying Deng, Marques Gilliam, Andres Moreno, Jordan McCammon, Naseera Bland, and Michael Penansky. Members of the Indicators Technical Teams and NCADAC Indicators Working Group are included in Kenney et al. (2014).

References

Binder CR, Hinkel J, Bots PW, et al. (2013) Comparison of frameworks for analyzing social-ecological systems. Ecol Soc 18(4):26

Buizer JL, Fleming P, Hays SL et al. (2013) Report on preparing the nation for change: building a sustained national climate assessment process. National Climate Assessment and Development Advisory Committee

Cloyd E, Moser SC, Maibach E, et al. (2015) Engagement in the third national climate assessment: commitment, capacity, and communication for impact. Clim Chang. doi:10.1007/s10584-015-1568-y

Cobb CW, Rixford C (1998) Lessons learned from the history of social indicators, Vol. 1: Redefining Progress San Francisco

Dorward A (2014) Livelisystems: a conceptual framework integrating social, ecosystem, development and evolutionary theory. Ecology and Society

Executive Office of the President (2013) The President's Climate Action Plan http://www.whitehouse.gov/sites/default/files/image/president27sclimateactionplan.pdf

Executive Order 13514 of October 5, 2009. Federal leadership in environmental, energy, and economic performance (Federal Adaptation Plans)

Executive Order 13642 of May 9, 2013. Making open and machine readable the new default for government information

Executive Order 13653 of November 1, 2013. Preparing the United States for the impacts of climate change (Section 4: Providing Information, Data, and Tools for Climate Change Preparedness and Resilience)

Fisher JA, Patenaude G, Meir P, et al. (2013) Strengthening conceptual foundations: analysing frameworks for ecosystem services and poverty alleviation research. Glob Environ Chang 23(5):1098–1111

Innes JE (1990) Knowledge and public policy: the search for meaningful indicators. Transaction Publishers, New Brunswick

Janetos AC, Kenney MA (2015) Developing better indicators to track climate impacts. Front Ecol Environ 13: 403–403. doi:10.1890/1540-9295-13.8.403

Janetos AC, Chen RS, Arndt D, et al. (2012) National climate assessment indicators: background, development, and examples. A technical input to the 2013 national climate assessment report. PNNL-21183, Pacific Northwest National Laboratory, Richland. pp. 59. http://Downloads.Globalchange.Gov/Nca/Technical_Inputs/NCA_Indicators_Technical_Input_Report.Pdf

Kenney MA, Janetos AC et al. (2014) National climate indicators system report. National Climate Assessment Development and Advisory Committee. http://www.globalchange.gov/sites/globalchange/files/Pilot-Indicator-System-Report_final.pdf

Lee KN (1993) Compass and gyroscope: integrating science and politics for the environment. Island Press, 255 pp

Lippman LH (2006) Indicators and indices of child well-being: a brief American history. Soc Indic Res 83:39–53

Meadow AM, Ferguson DB, Guido Z, et al. (2015) Moving toward the deliberate coproduction of climate science knowledge. Wea Climate Soc 7(2):179–191

Moser SC, Melillo J, Jacobs KL, et al. (2015) Aspirations and common tensions: larger lessons from the third US national climate assessment. Clim Chang. doi:10.1007/s10584-015-1530-z

Moss R, Scarlett PL, Kenney MA, et al. (2014) Ch. 26: Decision Support: Connecting Science, Risk Perception, and Decisions. Climate Change Impacts in the United States: The Third National Climate Assessment, Melillo JM, Richmond TC, and Yohe GW, eds., U.S. Global Change Research Program, 620–647. doi:10.7930/J0H12ZXG

National Research Council (1999) Nature's numbers: expanding the national economic accounts to include the environment. National Academies Press, Washington, DC

National Research Council (2000) Ecological indicators for the nation. National Academies Press, Washington, DC

National Research Council (2009a) Restructuring federal climate research to meet the challenges of climate change. National Academies Press, Washington, DC

National Research Council (2009b) Informing decisions in a changing climate. National Academies Press, Washington, DC

National Research Council (2013) Sustainability for the nation: resource connection and governance linkages. National Academies Press, Washington, DC

National Research Council (2015) Measuring progress toward sustainability: indicators and metrics for climate change and infrastructure vulnerability. Meeting in Brief, Roundtable on Science and Technology for Sustainability http://sites.nationalacademies.org/cs/groups/pgasite/documents/webpage/pga_168935.pdf

Ogburn WF (1933) Recent social trends in the United States. Report of the president's research committee on social trends. McGraw-Hill Book Company, New York

Steinemann A, Cavalcanti L (2006) Developing multiple indicators and triggers for drought plans. J Water Resour Plan Manag 132(3):164–174

The Heinz Center (2008) The State of the nation's ecosystems 2008: measuring the lands, waters, and living resources of the United States. The H. John Heinz III Center for Science, Economics and the Environment

USGCRP (2010) Ecosystem Responses to Climate Change: Selecting Indicators and Integrating Observational Networks. NCA Report Series, Vol. 5a. November 30-December 1, 2010, Washington, DC http://data.globalchange.gov/assets/68/23/2165abcb9561c165fb65ec2dcf3a/ecological-indicators-workshop-report.pdf

USGCRP (2011a) Societal Indicators for the National Climate Assessment. Kenney MA, Chen RS, et al. Eds. NCA Report Series, Vol. 5c. April 28–29, 2011a, Washington, DC. http://data.globalchange.gov/assets/f7/3f/d552dcec34ae9c23fcdefd61de65/Societal_Indicators_FINAL.pdf

USGCRP (2011b) Monitoring Climate Change and its Impacts: Physical Climate Indicators. NCA Report Series, Vol. 5b. March 29–30, 2011b, Washington, DC. http://data.globalchange.gov/assets/04/91/f875575999e1882ac421297f310b/physical-climate-indicators-report.pdf

USGCRP (2012) The National Global Change Research Plan 2012–2021: A Strategic Plan for the U.S. Global Change Research Program. USGCRP Report. http://data.globalchange.gov/assets/d0/67/6042585bce196769357e6501a78c/usgcrp-strategic-plan-2012.pdf

USGCRP (2015) National Climate Assessment & Development Advisory Committee (2011–2014) Meetings, decisions, and adopted documents. http://www.globalchange.gov/what-we-do/assessment/ncadac Accessed 14 Aug 2015

Victor DG, Kennel CF (2014) Climate policy: ditch the 2 °C warming goal. Nature 514:30–31. doi:10.1038/514030a

Walsh J, Wuebbles D, Hayhoe K et al. (2014) Ch. 2: Our Changing Climate. Climate Change Impacts in the United States: The Third National Climate Assessment, Melillo JM, Richmond TC, Yohe GW, Eds., U.S. Global Change Research Program, 19–67. doi:10.7930/J0KW5CXT

Weaver CP, Mooney S, Allen D, et al. (2014) From global change science to action with social sciences. Nat Clim Chang 4:656–659. doi:10.1038/nclimate2319

Climatic Change (2016) 135:97–109
DOI 10.1007/s10584-015-1547-3

Climate change impacts on ecosystems and ecosystem services in the United States: process and prospects for sustained assessment

Nancy B. Grimm[1] · Peter Groffman[2] · Michelle Staudinger[3] · Heather Tallis[4]

Received: 3 February 2015 / Accepted: 23 October 2015 / Published online: 4 November 2015

Abstract The third United States National Climate Assessment emphasized an evaluation of not just the impacts of climate change on species and ecosystems, but also the impacts of climate change on the benefits that people derive from nature, known as ecosystem services. The ecosystems, biodiversity, and ecosystem services component of the assessment largely drew upon the findings of a transdisciplinary workshop aimed at developing technical input for the assessment, involving participants from diverse sectors. A small author team distilled and synthesized this and hundreds of other technical input to develop the key findings of the assessment. The process of developing and ranking key findings hinged on identifying impacts that had particular, demonstrable effects on the U.S. public via changes in national ecosystem services. Findings showed that ecosystem services are threatened by the impacts of climate change on water supplies, species distributions and phenology, as well as multiple assaults on ecosystem integrity that, when compounded by climate change, reduce the capacity of ecosystems to buffer against extreme events. As ecosystems change, such benefits as water sustainability and protection from storms that are afforded by intact ecosystems are projected to decline across the continent due to climate change. An ongoing, sustained assessment that focuses on the co-production of actionable climate science will allow scientists from a range of disciplines to ascertain the capability of their forecasting models to project environmental and

This article is part of a special issue on "The National Climate Assessment: Innovations in Science and Engagement" edited by Katharine Jacobs, Susanne Moser, and James Buizer.

Electronic supplementary material The online version of this article (doi:10.1007/s10584-015-1547-3) contains supplementary material, which is available to authorized users.

✉ Nancy B. Grimm
nbgrimm@asu.edu

1 School of Life Sciences, Arizona State University, Tempe, AZ, USA

2 Cary Institute of Ecosystem Studies, Millbrook, NY, USA

3 Northeast Climate Science Center, Amherst, MA, USA

4 The Nature Conservancy, Santa Cruz, CA, USA

ecological change and link it to ecosystem services; additionally, an iterative process of evaluation, development of management strategies, monitoring, and reevaluation will increase the applicability and usability of the science by the U.S. public.

1 Introduction

Ecosystems are the basic components of the Earth's surface, containing both living organisms and the non-living matrix of air, soil, fresh water, ocean waters, sediments, and geologic parent material. The tremendous variety of ecosystems supports an astounding diversity of life, including the human species. Impacts of climate change on ecosystems often translate directly or indirectly to impacts on people (Fig. 1). The third National Climate Assessment (NCA) explicitly recognized that relationship, couching its consideration of ecosystem impacts in the context of how climate change in the United States affects the benefits ecosystems provide to individuals, communities, societies, and social institutions (hereafter, simply, 'people'; Groffman et al. 2014). Ecologists refer to the benefits that people derive from ecosystems as ecosystem services (Daily 1997; MEA 2005). Previous U.S. NCAs (National Assessment Synthesis Team 2000; Karl et al. 2009) projected that ecosystem services would likely decline because of multiple observed and projected changes in species and ecosystems. However, in this third NCA, key findings were explicitly ranked and evaluated in terms of their impacts on people.

Climate assessments may provide comprehensive evaluations of complex interactions among regions, sectors, and other landscape-scale stressors such as land-use and land-cover changes. Sectors for the United States as defined in the NCA include water resources, energy, transportation, agriculture, forestry, and human health, as well as ecosystems and biodiversity. The third NCA also explicitly considered relationships among species, ecosystems, and ecosystem services; forestry, fish, and wildlife resources; water, energy, and land use; land use and human population dynamics and their interactions with climate change; biogeochemical cycles and climate change; and urban systems, infrastructure, and human vulnerability. Thus, it was a more integrated assessment than the previous two, with many findings shared among different chapters and sectors. It also was the first assessment to deliberately and

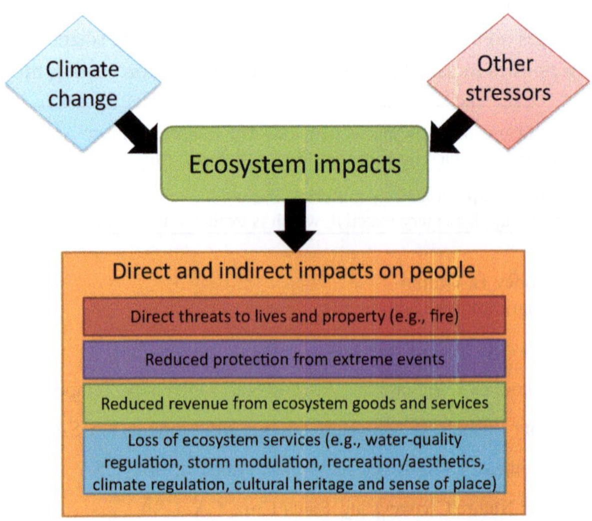

Fig. 1 Direct and indirect impacts of climate change and other, confounding stressors on people, mediated through influences on ecosystems

systematically examine how people in different settings within the United States (urban, rural, and tribal lands) are experiencing climate change and will continue to do so in the future (Melillo et al. 2014).

1.1 Objectives

As part of this special issue on the processes and outcomes of the third NCA, this article discusses how the ecosystems, biodiversity, and ecosystem services assessment was developed, how it differed from the previous NCAs, specifically in terms of its treatment of ecosystem services, and how it may serve as a model for future assessments through collaborative and collegial modes of production (Meadow et al. 2015). As with the ecosystems chapter of the third NCA, the impacts discussed are specific to the United States and are not meant to cover global ecosystems and ecosystem services.

The article summarizes the process, findings, and benefits of creating a chapter of the NCA on ecosystems that emphasizes ecosystem services through a transdisciplinary research approach. Both development of technical input and synthesis of technical input and recent literature to develop the final chapter are described. Two intermediate sections discuss findings in the context of an ecosystem services perspective and provide a case study for how a transdisciplinary approach can be used to address complex and socially relevant scientific problems (i.e., "wicked problems") (Meadow et al. 2015). The article concludes with recommendations from lessons learned on how best to organize and carry out an integrative assessment with findings that are accessible and useable by stakeholders, which in this case were the President, the Congress, and the people of the United States.

1.2 Rationale for an ecosystem services perspective

In the five years since the last NCA report was issued (Karl et al. 2009), we have seen increasingly frequent reports of weather-related natural disasters (Kunkel et al. 2010; e.g., Hurricanes Irene and Sandy in the U.S.) and a dramatic rise in expenditures for emergency management efforts in response to wildfires, hurricanes and other storms, heat waves, and sea surges (e.g., The Royal Society 2014). An understanding that ecosystems can provide protection from some extreme events has spurred interest in such regulating ecosystem services. For example, in response to damage from "superstorm" Sandy in October 2012, modifications of shorelines in the greater New York City region have included expanding and restoring ecosystem features such as salt marsh habitat, as an alternative or accompaniment to 'hard' or engineered infrastructure. Because housing developments in the region that had razed dunes were much harder hit during Hurricane Sandy than those that left dunes intact (Foderaro 2012), there is renewed interest in dune restoration (Cutter et al. 2014). A second example of protection from extreme events was the Las Conchas fire in New Mexico in 2011, where complex interactions among forest health, fires, urbanization, erosion in heavy rainstorms, and impacts on lives, structures, and water supplies were regulated by the ecosystem services of soil stability, water-quality, and water provision (Grimm et al. 2013b). Examples like these show the direct relevance of healthy ecosystems to people and reinforce our assertion that the ecosystems services lens used in the chapter on ecosystems, biodiversity, and ecosystem services (hereafter, the 'ecosystems' chapter) is a valuable and appropriate way to assess climate-change impacts across most regions and some sectors.

Since its introduction in the 1990's (Daily 1997), the concept of ecosystem goods and services has taken hold in both the ecological research community and the resource management community. Using this concept in the third NCA prioritized ecosystem changes that had demonstrable impacts on people. But consideration of the goods and services provided by ecosystems is not unique to this third NCA: previous United States assessments have observed that the impacts of climate change on ecosystems are likely to extend to society because people depend on ecosystems to provide those benefits. Both the first and second NCA reports point out the difficulty of quantifying such benefits, and both aver that, while some services will be enhanced by climate change, most will decline (NAST 2000, p 24; Karl et al. 2009, p 87). Indeed, the foundational Strategic Assessment Report for ecosystems points out, "It is possible to make some generalizations from the literature on the physical changes in ecosystems, but interpreting what these changes mean for services provided by ecosystems is very challenging…" (Backlund et al. 2008). Although linking ecosystem change to specific ecosystem services remains difficult, the third NCA took the approach of ranking and evaluating impacts (key findings) based on specific human outcomes. Threats to water availability and quality from ecosystem change and harmful algal blooms, people and property affected by sea level rise, and loss of life and property associated with fires are cited (Groffman et al. 2014). For example, the number of U.S. counties at risk of water shortage was projected to increase to 32 % in 2050 from 10 % today.

Several recent international and national assessments also have demonstrated the relationships among biodiversity, ecosystems, and ecosystem services. These efforts were intended to increase awareness by policy makers and the general public of the value that ecological systems provide both directly (e.g., food, timber) and indirectly (e.g., supportive services such as nutrient and water cycling) to society, as well as the risks that their degradation will cause if left unchecked. One effort, the Intergovernmental Science-Policy Platform on Biodiversity and Ecosystem Services (IPBES), which stemmed from the recommendations of the Millennium Ecosystem Assessment (MEA 2005) and the International Mechanism of Scientific Expertise on Biodiversity (2007), was convened to serve as platform similar to the Intergovernmental Panel on Climate Change (IPCC), but with a focus on synthesizing, assessing, and critically evaluating the global condition of biodiversity, ecosystems, and ecosystem services (Perrings et al. 2011). Independent scientific assessments conducted through IPBES are expected to increase the capacity of the Convention on Biological Diversity and participating nations of IPBES to make informed decisions about the management and conservation of their natural resources. Another example is the UK National Ecosystem Assessment (UK NEA 2011, Mace et al. 2012), which used an ecosystem services framework to increase understanding and awareness of the economic and societal values of natural resources, and promote the development of tools and products to inform decision-making and adaptive management of ecological systems. The NCA ecosystems chapter was inspired by these prior efforts and drew from their experience in developing its focus.

2 The process: assessment of present and future ecosystem impacts

As with the other chapters of the NCA, the ecosystems chapter was developed by distilling and synthesizing extensive technical input from the broader scientific community. What set the ecosystems chapter apart from other working groups was the transdisciplinary expert team, interactive process, and resources devoted to generate findings. Here we highlight the participation, process, and lessons learned from two phases of the chapter's development as a model for future assessments seeking to address complex and multi-dimensional areas of scientific knowledge. Production of both a large technical input report and the report chapter itself raised questions about how best to meet the challenges posed by each step; these questions are answered, in part, with the final section on recommendations. An assessment represents a synthesis intended to provide potential answers to policy-relevant questions (Grimm et al. 2013b); we couch these questions in terms of how the assessment can be continually updated, made accessible and actionable to diverse stakeholders.

2.1 Development of a technical input report

In July 2011, the NCA Development and Advisory Committee (NCADAC), via the U.S. Global Change Research Program (USGCRP), issued a "Request for Information" in the Federal Register (Fig. 2), resulting in hundreds of technical input items received across all sectors, with 128 directed to the ecosystems chapter. The most important contribution was the Biodiversity, Ecosystems, and Ecosystem Services report (Staudinger et al. 2012), which is discussed further in this paper.

The USGCRP encouraged agency participation in developing technical input through personal contacts and the Interagency NCA working group. Scientists at the US Geological Survey's (USGS) National Climate Change and Wildlife Science Center (NCCWSC) had emphasized ecosystem services in several recent initiatives and were actively engaged with the IPBES; in collaboration with NCADAC and NCA staff, NCCWSC supported the production of a technical input report and workshop for biodiversity, ecosystems, and ecosystem services, including supporting a postdoctoral fellow dedicated to the effort.

The development of the *2012 Technical Input to the NCA on Impacts of Climate Change on Biodiversity, Ecosystems, and Ecosystem Services* (Staudinger et al. 2012; referred hereafter as

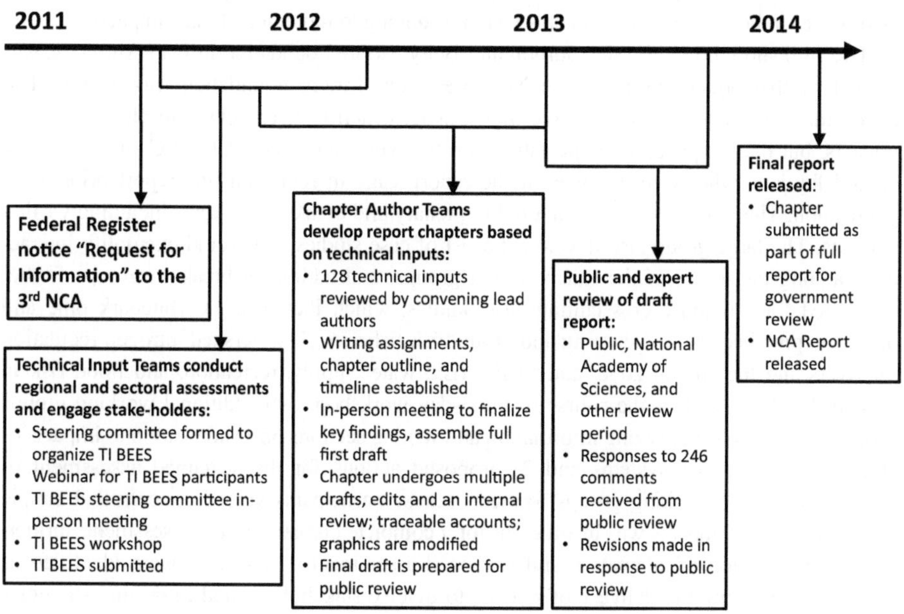

Fig. 2 Process and timeline to develop the ecosystems chapter of the third National Climate Assessment

the TI BEES) thus began through these collaborative and collegial interactions. A steering committee was formed for the TI BEES, with members from academia, nongovernmental organizations, U.S. government agencies (including a NCA staff member), and the NCADAC. The steering committee outlined preliminary objectives, topics, and integrative case studies to be included in the TI BEES, reviewed previous assessments upon which the TI BEES could build, and secured funding to support an in-person steering committee meeting and the proposed workshop. Perhaps most importantly, they compiled and invited a list of experts and stakeholders, drawn from a broad range of federal and state agencies, academic institutions, and non-governmental organizations, to develop the TI BEES. Selection of the TI BEES team was a critical step toward finding a balance of interdisciplinary expert knowledge that spanned from academia to conservation to management and policy, and had wide geographic representation from the U.S. The targeted experts would have to commit to meeting an aggressive timeline to produce a product whose guidelines were simultaneously being developed and refined. An informational webinar was held in October 2011 for potential participants, which provided an overview of past assessments, the NCA process, expectations for their participation, and the path forward for the TI BEES.

Approximately sixty experts committed to the production of the TI BEES and participated in one of four working groups that considered climate-change impacts on 1) biodiversity, 2) ecosystems, 3) ecosystem services as well as 4) other (non-climate) stressors. Working groups were assembled based on a pre-webinar survey asking for expertise and interest in relevant topics. Each workgroup met separately by phone to outline content for their sections and assign writing tasks prior to meeting in person at the workshop in January 2012. An example of an approach used by some working groups to synthesize the best available science and support the development of the TI BEES was to identify (primarily from literature published since the previous NCA in 2009) the top 10 most influential and groundbreaking papers on each subject (e.g., climate impacts on biodiversity); then, through an iterative process, they extracted the key scientific advances within each topic area over recent years and identified major gaps in knowledge, as well as surprises in how species and systems were responding to mounting climate impacts.

The workshop, hosted by the Gordon and Betty Moore Foundation in Palo Alto, CA, was funded by that organization and the NCCWSC. The Moore Foundation was interested in contributing to work that would have high impact in terms of responses to climate change impacts, but they imposed no expectations on the workshop. The NCCWSC similarly were neutral funders, who were involved at the report stage in reviewing the report prior to its publication. The workshop was structured in plenary and breakout sessions, including writing sessions. The large group also developed a list of case studies that would exemplify interactions among the topics of the four working groups, and additional breakout groups further developed four of those crosscutting case studies, which included: 1) whitebark pine and mountain pine beetles; 2) New England groundfish fisheries; 3) watershed nitrogen regulation for rivers and the coastal zone; and 4) fire risk, water supply, recreation, and flood risk in western US forests. Over the course of the 3-day workshop, two additional working groups formed and began synthesizing information and writing sections on 1) adaptation to impacts of climate change on ecosystems and 2) proposed actions for the sustained assessment of ecosystems. These working groups formed because participants voiced concern that simple documentation of impacts was insufficient for a complete climate change assessment; furthermore, the inclusion of adaptation and sustained assessment topics areas were identified as critical to advancing the delivery of science to diverse stakeholders and informing strategies and actions to respond to climate impacts on ecological systems.

The development of the TI BEES was a rapid process, conducted over approximately a 6-month period (Fig. 2). Overall, the process represents a successful model for synthesizing a vast array of scientific information in short order, yet questions remain from the experience. How could the information collection process be expanded and streamlined to ensure comparable quality while reducing individual effort in future assessments? Was the participation of a large and diverse group of experts valuable, and what mechanisms could attract them and reward their continued participation? Finally, what were the financial costs with the technical input process and were they appropriate?

Key to increasing the efficiency of information gathering in future assessments would be the establishment of an information database equipped with mechanisms to extract and continually update information from contributors. Such information would include summaries of scientific discovery, case studies of ecosystem responses to climate change impacts, examples of management and conservation tools that have been developed and implemented, and lessons learned from the successes and failures of those efforts. Contributors to the assessment process would be encouraged to update scientific and management information as it is produced.

Keeping a large and diverse group of experts engaged ensures that procured information is relevant to a wide range of audiences and includes as much of the available evidence as possible. Greater transparency about the NCA process is needed, however. We noted that there was much confusion among the participants (both for this and other technical input reports) about exactly what they were contributing to and how it might be used to produce the actual NCA report. The informational webinar attempted to speak to the organization and goals of the effort, but many aspects of the process and overall direction of the NCA and the technical input were being developed in parallel, making it challenging to communicate precisely what participants were committing to. Communication with experts about exactly how their contributions will be used and what kind of information is most useful will be essential to maintaining engagement in the context of a sustained assessment. If the objective is to encourage broad input from a diversity of sectors, posting notices in the Federal Register, although necessary, may not be sufficient. Instead, wide-reaching communication vehicles like NCANet (http://ncanet.usgcrp.gov/) may allow access to a broader group of potential contributors.

Finally, producing the technical input report did incur a substantial cost: travel for the steering committee and sixty workshop participants; food, housing, and meeting facilities at the workshop; and the post-doc salary. We aver that the in-person meetings were invaluable to ensuring that the TI BEES included diverse perspectives. The workshop was deliberately structured to be interactive to maximize idea exchange over the course of the meeting. Through this model, the workshop promoted engagement, communication, and relationship building among participants. The steering committee was able to rapidly coordinate and refine content and objectives—examples including the development of case studies and the addition of two working groups—to meet the broader goals of the overall report. Such responsiveness to evolving content could not have been accomplished through a virtual meeting format.

A post-doctoral fellow dedicated to workshop planning and report preparation was the sole member of the steering committee and working group teams whose responsibilities and time were completely devoted to the effort. For a group and topic area this broad in scope, it was essential to have such a point person to effectively organize and coordinate the content, teams, and meetings, especially given the aggressive timelines of the third NCA.

In addition to the full report, which is available to the public on the Globalchange.gov website, the technical input workgroup also produced an open-access special issue in the highly ranked Ecological Society of America journal, *Frontiers in Ecology and the*

Environment (Grimm et al., 2013a, b, Grimm and Jacobs 2013, Nelson et al. 2013, Staudinger et al. 2013, Staudt et al. 2013, Stein et al. 2013). Through these products, coupled with ecosystems chapter of the NCA report itself, the current understanding of the impacts of climate change on biodiversity, ecosystems, and ecosystem services in the United States was communicated and made freely available to multiple audiences (e.g., the public, Congress, scientists), amplifying the potential impact of the entire effort.

2.2 Development of the report chapter

As with all NCA chapters, authors who were considered well-regarded scientists in the subject area were selected and invited to develop the ecosystems chapter by the NCA Development and Advisory Committee (DAC). The two convening lead authors, an academic scientist and a scientist from a leading conservation NGO, were joined by four university faculty members and a research institute scientist specializing in different areas of ecology and conservation biology, one US government employee, and one academic working with the Natural Capital Project, which functioned as a boundary organization. The author team began work in March 2012, completed a first draft by June 2012, and made revisions in response to the rigorous review prepared by the NCA staff in consultation with convening lead authors. The final chapter (Groffman et al. 2014) was completed in February 2014, following a formal public and government review (Fig. 2).

A major challenge to the development of the chapter was meeting the NCADAC-imposed length restriction of eight pages of text. To put this in perspective, the author team needed to consider and digest over 1000 pages from 128 technical inputs, including the 296-page TI BEES report, to decide upon the key findings of the chapter. Originally, the author team envisioned the chapter organized around several 'spheres' reflecting different ecological impacts: the dry, the wet, the cold, and "who's there?" Next, the team decided that "who's there?"—a consideration about how individual species within ecosystems have been and will be affected by climate change—should be integrated into the other spheres. Later, this differentiation was abandoned in favor of highlighting the ecosystem changes that are likely to be of greatest impact in terms of ecosystem services. The winnowing of ideas was strongly driven by the NCA guidelines requiring a short chapter, with a small number of key findings. The development and imposition of these guidelines aided in producing a concise and accessible amount of information that highlighted genuinely high-impact key findings.

Many of the main findings about ecosystems and species translated directly into key findings for ecosystem services. As such, framing the findings through the lens of "what matters most to people" integrated broad topic areas and packaged them in a way that was meaningful to the policy and public audiences of the assessment, as well as making them actionable to the natural resource management community (Mace et al. 2012). The author team also considered the potential synergy of the ecosystem chapter's key findings in relationship to other sectors within the assessment (e.g., water, agriculture, coastal, and forestry chapters) that would potentially address topics relevant to ecosystems, species, and ecosystem services.

The diversity of perspectives and disciplines captured within the ecosystems team as well as the process used to develop the chapter, were transdisciplinary in nature, thus leading to key findings that effectively communicated an enormous amount of material in a way that was accessible and practical to a wide range of audiences and stakeholders (Meadow et al. 2015). Many conversations within the team focused on following a thread from a noted change in a species and/or process, through to impacts on an ecosystem, and finally to the relevance of those impacts for people, with different authors weighing in at each phase. For example, academic knowledge of species changes and how they might affect ecosystems would be coupled with experience from Nature Conservancy and Natural Capital projects translating changes in ecosystem services to monetary or non-monetary impacts on people. This type of iterative discussion generated a shared understanding of which findings reflected strong connections across components of systems. The mix of expertise also helped translate complex and large amounts of information into findings summarized in concise, clear points that were understandable to both experts and non-experts. For example, a non-expert can sometimes be the best filter, identifying the most critical findings from the complexity and detail held so dear by experts. The findings and text also evolved as a result of the diverse professions represented, ending in a document that expressed scientific rigor, detail, and certainty, while also addressing the practical needs for clarity and relevance to policy and action. Relevance was judged based on the authors' knowledge of policy concerns but might have profited from greater involvement at this stage by practitioners.

Several rounds of agency, community, National Research Council, and NCA internal reviews yielded 246 comments that required response. Review comments were generally supportive of the ecosystem services framing, and focused largely on language and separating climate impacts from other stressors. The convening lead authors worked with NCA staff to resolve comments on the chapter, which was a time-consuming process with constantly shifting guidelines and requirements. The multi-phased review and revision process was arduous but necessary and important, as it provided stakeholders with another opportunity within the assessment process to contribute and shape the overall messages. For example, the review identified some inaccuracies and missing pieces in the chapter, such as the unintentional omission of information on species extinctions linked to climate change within the biodiversity component of the chapter. Ultimately, the review and revision process led to a final product that reflects the needs of the targeted audience and stakeholders.

How could the entire process—from technical input to chapter development— be improved for future assessments? Based on our experience, having technical input and chapter teams share some members increased efficiency, especially given the short time available to the author team for chapter development. Numerous benefits in terms of relating ecosystem impacts to people accrued from the diverse experiences of the author team and the practice of inviting multisectoral authors; this helped amplify the relevance, breadth, and accessibility of assessment findings. Alternatively, clear and firm guidelines and a timeline for the multiple phases of assessment, review, and revision would be helpful, so that authors and support staff are not continually asked to alter their approach or make revisions on short notice. Developing the NCA as on ongoing, iterative process would almost certainly help to bring about these changes.

3 Findings in the context of the ecosystem services perspective

Biodiversity and ecosystems are shifting in time and space as they respond to complex climate (e.g., temperature, water), biological (e.g., trophic interactions), and anthropogenic (e.g., land-use change) drivers worldwide (IPCC 2014) and in the US specifically (Grimm et al. 2013b; Groffman et al. 2014; Staudinger et al. 2013). Characterizing individual and localized responses is critically important because it identifies particular vulnerabilities of species and ecosystems, as well as adaptive capacity to cope with ongoing climate changes. Which species are adapting and which are declining will have direct impacts on people through changes in available resources. Already, fishermen adjust where and when they cast their nets and which species they target (Pinsky and Fogarty 2012). Insect pests are taking advantage of warmer winter conditions in many parts of the U.S. and are thereby changing forest landscapes, including tree species that provide timber, income, and a way of life for communities and tribal nations (Bennett et al. 2014, Joyce et al. 2014). Forest thinning, a common management practice, raises the risk for drought impacts, which affects ecosystem services of forests (D'Amato et al. 2013). The connections between impacts of climate change on ecological systems and consequences for the goods and services that people depend on must be elucidated and communicated. Contributors to the ecosystems chapter strived to frame the impacts of climate change on biodiversity and ecosystems in the context of opportunities and losses that would resonate with the U.S. public and decision makers. Although ecosystem-based approaches to conservation and management of natural resources are not new concepts, this is the first time that an ecosystem services approach has been a primary emphasis in the NCA. As discussed earlier, the ecosystem services context also is prevalent among international assessments seeking to strengthen the influence of their key messages across a range of scientific, management, and public audiences.

As conservation and management practioners approach the task of anticipating and responding to the impacts of climate change on natural resources through adaptation approaches, they increasingly will need to contemplate landscape scales that consider not only where species and biomes have been historically but where they have the greatest likelihood of being in the future (Stein et al. 2013). This includes, for example, working with communities to restore coastal habitats, such as salt marshes and dunes, that provide protection against extreme weather events, while also enabling species movements inland through corridors, as sea-level rise and coastal storms erode or inundate coastlines. Those developing strategies for adaptation to climate change will need to engage stakeholders early and often, throughout both inception and implementation stages, to ensure that community needs and values are incorporated into the overall strategy and balanced with sustaining ecological targets.

4 Opportunities for ecologists afforded by the assessment

Perhaps the most compelling scientific aspect of the NCA process was the opportunity to evaluate the accumulated evidence, or lack thereof, in support of past key NCA projections regarding climate impacts on ecosystems, biodiversity, and ecosystem services. Many of the key messages of the first two NCAs are now being realized; however, in certain cases our understanding is becoming more refined while in others there are surprises. Where have we been right, where have we been wrong, and what can we learn from where we have been wrong? For example, ecologists have long projected that forests would grow better in a warmer, wetter world with higher levels of atmospheric CO_2. Indeed, trees *are* growing better all across the continent (Williams et al. 2012)—except where they are not. Non-conforming regions include vast areas of dead and declining conifer forests in the West (van Mantgem et al. 2009) and vanishing hemlock stands in the East (Paradis et al. 2008). The loss of extreme low winter temperatures and extended growing seasons have facilitated the spread and/or persistence of insect pests that can significantly damage forests (Grimm et al. 2013a); a native pest outbreak (pine bark beetle; e.g., Bentz et al. 2010) is the culprit, while in the East, loss of low temperature has facilitated the spread of an invasive pest, the Hemlock wooly adelgid (Paradis et al. 2008). These examples demonstrate regional or local exceptions to broader trends.

Taking a continental-scale view, such as is required for an assessment of impacts across the United States, allows ecologists to examine where we were wrong in our earliest projections, to determine if aspects of climate change underlie these unexpected results, and to identify poorly understood mechanisms that need further investigation. The dying forests example illustrates the complexity of climate-change effects on ecosystems, triggers and thresholds that may result in opposite outcomes in different locations, as well as the need to integrate expertise from multiple disciplines (e.g., climatology, forestry, and entomology in this case). Thus, an ongoing, iterative process of prediction, evaluation, and revision of ideas will greatly improve climate-change science. Furthermore, assessment that incorporates unique regional responses will enhance the ability of managers to anticipate and respond to climate-change effects in ways that are appropriate to place.

5 Recommendations

The experiences of the authors of this paper as members of technical input and author teams for the ecosystems chapter lead us to conclude this paper with a set of recommendations, offered in the hope that they will have relevance and value to future NCAs.

- Future NCAs should retain and further develop the focus on ecosystem services, as the impacts of climate change on people are often the most relevant and of the greatest concern to target audiences. Many impacts on ecosystems may be translated to impacts on ecosystem services, and thereby, on people.
- Technical input from the scientific community in diverse sectors is essential but must be streamlined by providing infrastructure and mechanisms to enable a continuous flow of information from contributors, between and throughout assessments. Resources must be devoted to the accumulation and evaluation of scientific information as it is produced, such as staff dedicated to communicating and seeking out new information and to keeping databases up to date in anticipation of each next assessment. The outstanding NCA website should be expanded to allow this flow of information.
- Technical input from diverse stakeholders, including managers, legislators, and practitioners, is essential at all stages of the process to ensure that the science synthesis and findings are salient and useable for management, policy, and action. As described here, interfacing with all of these groups is time consuming; therefore, to do so comprehensively may require extending outreach and feedback timelines within the different phases of the NCA (e.g., review and revision stages).
- Incentives for contributors, in terms of both resources to enable their participation and clear expectations for the fate of their contributions, would improve the acquisition of technical input. Limited funding across all federal agencies will be an obvious barrier to overcome in making incentives a reality; furthermore, a compressed timeline for producing the NCA may preclude substantial involvement of many potential contributors. More clear and firm timelines for the review and revision process would make it easier to recruit and retain authors for future assessments. Developing the NCA as on ongoing, iterative process would almost certainly help to bring about these changes.
- In person, face-to-face meetings remain an essential way to stimulate the exchange of ideas and good syntheses of available knowledge for application to assessments and are well worth the financial investment, especially for broad topic areas.

- A survey of TI BEES and chapter authors could be initiated to estimate the total number of hours spent on development of final products, ascertain preferences for engagement methodology (e.g., virtual vs. in person meetings), and solicit opinions on how the process could be improved.
- The transdisciplinary approach involving multiple, crosscutting sectors and topics is a strength of this NCA and should be a model for future assessments; however, a means for promoting better communication and integration among related topics is needed. Dedication of a point person with responsibility to interface with other sectoral teams is one possibility.
- There is a strong need for the NCA to be an ongoing iterative process of making projections, evaluating the literature across disciplines and regions, and revising ideas, as this will greatly improve climate-change science and the ability of the U.S. to anticipate and respond to climate-change impacts.

Acknowledgments The authors acknowledge the contributions of many scientists and other stakeholders to the production of the TI BEES and the Ecosystems Chapter. We are especially grateful to TI BEES steering committee members Peter Kareiva and Shawn Carter and author team members Carter, Kareiva, Josh Lawler, Michelle Mack, and Virginia Matzek. NCA staff member Susan Aragon-Long was tireless in her efforts during the review and revision process. We acknowledge the work of reviewers, whose comments helped to improve this paper.

References

Backlund P, Janetos A, Schimel DS, Hatfield J, Ryan MG, Archer SR, Lettenmaier D (2008) Executive summary. The effects of climate change on agriculture, land resources, water resources, and biodiversity in the United States. A Report by the U.S. Climate Change Science Program and the Subcommittee on Global Change Research. Washington, DC., USA, In, 362 pp

Bennett, T. M. B., N. G. Maynard, P. Cochran, R. Gough, K. Lynn, J. Maldonado, G. Voggesser, S. Wotkyns, and K. Cozzetto. 2014. Ch. 12: Indigenous Peoples, Lands, and Resources. Climate Change Impacts in the United States: The Third National Climate Assessment, J. M. Melillo, Terese (T.C.) Richmond, and G. W. Yohe, Eds., U.S. Global Change Research Program, 297–317. doi: 10.7930/J09G5R1

Bentz BJ, Regniere J, Fettig CJ, et al. (2010) Climate change and bark beetles of the western United States and Canada: direct and indirect effects. Bioscience 60:602–613

Cutter, S. L., W. Solecki, N. Bragado, J. Carmin, M. Fragkias, M. Ruth, and T. J. Wilbanks. 2014. Ch. 11: Urban systems, infrastructure, and vulnerability. Pages 282–296 *in Climate Change Impacts in the United States: The Third National Climate Assessment*. J. M. Melillo, Terese (T.C.) Richmond, and G. W. Yohe, Eds., U.S. Global Change Research Program. doi: 10.7930/J0F769GR

Daily, GC (1997) Nature's services: societal dependence on natural ecosystems. Island Press, Washington, DC

D'Amato AW, Bradford JB, Fraver S, Palik Brian J. BJ (2013) Effects of thinning on drought vulnerability and climate response in north temperate forest ecosystems. Ecol Appl 23:1735–1742

Groffman, P. M., P. Kareiva, S. Carter, N. B. Grimm, J. Lawler, M. Mack, V. Matzek, and H. Tallis. 2014. Ch. 8: Ecosystems, Biodiversity, and Ecosystem Services. Climate Change Impacts in the United States: The Third National Climate Assessment, J. M. Melillo, Terese (T.C.) Richmond, and G. W. Yohe, Eds., U.S. Global Change Research Program, 195–219, doi: 10.7930/J0TD9V7H.

Grimm NB, Chapin FS, Bierwagen B, Gonzalez P, Groffman PM, Luo Y, Melton F, Nadelhoffer K, Pairis A, Raymond PA, Schimel J, Williamson CE (2013a) The impacts of climate change on ecosystem structure and function. Front Ecol Environ 11:474–482

Grimm NB, Jacobs KL (2013) Evaluating climate impacts on people and ecosystems. Front Ecol Environ 11: 455. doi: 10.1890/1540-9295-11.9.455

Grimm NB, Staudinger MD, Staudt A, Carter SL, Chapin FS, Kareiva P, Ruckelshaus M, Stein BA (2013b) Climate-change impacts on ecological systems: introduction to a US assessment. Front Ecol Environ 11: 456–464

Foderaro, L. W. November 20, 2012. "After Saving Fire Island, Line of Defense Is Lost to the Sea". *The New York Times*. p. A18. Retrieved 10/2/2014.

Joyce, L. A., S. W. Running, D. D. Breshears, V. H. Dale, R. W. Malmsheimer, R. N. Sampson, B. Sohngen, and C. W. Woodall. 2014. Ch. 7: Forests. *Climate Change Impacts in the United States: The Third National Climate Assessment*, J. M. Melillo, Terese (T.C.) Richmond, and G. W. Yohe, Eds., U.S. Global Change Research Program, 175–194. doi: 10.7930/J0Z60KZC

Karl TR, Melillo JM, Peterson TC (2009) Global climate change impacts in the United States. Cambridge University Press

Kunkel, K. E., D. R. Easterling, D. A. R. Kristovich, B. Gleason, L. Stoecker, and R. Smith. 2010. Recent increases in U.S. heavy precipitation associated with tropical cyclones. *Geophysical Research Letters*: 37. doi: 10.1029/2010gl045164

Mace GM, Norris K, Fitter AH (2012) Biodiversity and ecosystem services: a multilayered relationship. Trends in Ecology and Evolution 27:19–26

MEA, M. E. A (2005) Ecosystems and human well-being: biodiversity synthesis. World Resources Institute, Washington, D.C.

Meadow AM, Ferguson DB, Guido Z, Horangic A, Owen G, Wall T (2015) Moving toward the deliberate coproduction of climate science knowledge. Weather, Climate, and Society 7:179–191

Melillo JM, Richmond T(TC), Yohe GW (eds) (2014) Climate change impacts in the United States: The third national climate assessment. U.S. Global Change Research Program, 841 pp. doi: 10.7930/J0Z31WJ2

National Assessment Synthesis Team (2000) Climate change impacts on the United States: The potential consequences of climate variability and change. US Global Change Research Program, Washington DC

Nelson EJ, Kareiva P, Ruckelshaus M, Arkema K, Geller G, Girvetz E, Goodrich D, Matzek V, Pinsky M, Reid W, Saunders M, Semmens D, Tallis H (2013) Climate change's impact on key ecosystem services and the human well-being they support in the US. Front Ecol Environ 11:483–893. doi: 10.1890/120312

Paradis A, Elkinton J, Hayhoe K, Buonaccorsi J (2008) Role of winter temperature and climate change on the survival and future range expansion of the hemlock woolly adelgid (*adelges tsugae*) in eastern North America. Mitig Adapt Strateg Glob Chang 13:541–554

Perrings C, Duraiappah A, Larigauderie A, Mooney H (2011) The biodiversity and ecosystem services science-policy interface. Science 331:1139–1140

Pinsky ML, Fogarty M (2012) Lagged socio-ecological responses to climate and range shifts in fisheries. Clim Chang 115:883–891

The Royal Society. 2014. Resilience to Extreme wEather The Royal Society Policy Report, DES3400, ISBN: 978–1-78252-113-6

Staudinger, M.D., Grimm, N., Staudt, A., Carter, S., Chapin, F.S.III, Kareiva, P., Ruckelshaus, M., and B. Stein. 2012. Impacts of Climate Change on Biodiversity, Ecosystems, and Ecosystem Services: Technical Input to the 2013 National Climate Assessment. Cooperative Report to the 2013 National Climate Assessment. 296 p. Available at: http://assessment.globalchange.gov

Staudinger MD, Carter S, Cross M, Dubois N, Duffy E, Enquist C, Griffis R, Hellmann J, Lawler J, Morrison S, O'Leary J, Stein B, Turner W (2013) Biodiversity in a changing climate: a synthesis of current and projected trends in the United States. Front Ecol Environ 11:465–473

Staudt A, Leidner AK, Howard J, Aa Brauma K, Dukes JS, Hansen LJ, Paukert C, Sabo J, Solórzano LSA (2013) The added complications of climate change: understanding and managing biodiversity and ecosystems. Front Ecol Environ 11:494–501. doi: 10.1890/120275

Stein BA, Staudt A, Cross MS, Dubois NS, Enquist C, Griffis R, Hansen LJ, Hellmann JJ, Lawler JJ, Nelson EJ, Pairis A (2013) Preparing for and managing change: climate adaptation for biodiversity and ecosystems. Front Ecol Environ 11:502–510

UK National Ecosystem Assessment (2011) The UK national ecosystem assessment: synthesis of the Key findings. UNEP-WCMC, LWEC, UK

Van Mantgem PJ, Stephenson NL, Byrne JC, Daniels LD, Franklin JF, Fule PZ, Harmon ME, Larson AJ, Smith JM, Taylor AH, Veblen TT (2009) Widespread increase of tree mortality rates in the western United States. Science 323:521–524

Williams CA, Collatz GJ, Masek J, Goward SN (2012) Carbon consequences of forest disturbance and recovery across the conterminous United States. Glob Biogeochem Cycles 26:GB1005. doi: 10.1029/2010GB003947

Climatic Change (2016) 135:111–126
DOI 10.1007/s10584-015-1535-7

Engagement with indigenous peoples and honoring traditional knowledge systems

Julie Maldonado[1] · T. M. Bull Bennett[2] · Karletta Chief[3] · Patricia Cochran[4] · Karen Cozzetto[5] · Bob Gough[6] · Margaret Hiza Redsteer[7] · Kathy Lynn[8] · Nancy Maynard[9] · Garrit Voggesser[10]

Received: 11 January 2015 / Accepted: 16 August 2015 / Published online: 26 October 2015

Abstract The organizers of the 2014 US National Climate Assessment (NCA) made a concerted effort to reach out to and collaborate with Indigenous peoples, resulting in the most comprehensive information to date on climate change impacts to Indigenous peoples in a US national assessment. Yet, there is still much room for improvement in assessment processes to ensure adequate recognition of Indigenous perspectives and Indigenous knowledge systems. This article discusses the process used in creating the Indigenous Peoples, Land, and Resources NCA chapter by a team comprised of tribal members, agencies, academics, and non-governmental organizations, who worked together to solicit, collect, and synthesize traditional knowledges and data from a diverse array of Indigenous communities across the US. It also

This article is part of a special issue on "The National Climate Assessment: Innovations in Science and Engagement" edited by Katharine Jacobs, Susanne Moser, and James Buizer

Electronic supplementary material The online version of this article (doi:10.1007/s10584-015-1535-7) contains supplementary material, which is available to authorized users.

✉ Julie Maldonado
jkmaldo@gmail.com

1 Livelihoods Knowledge Network, Santa Barbara, CA, USA

2 Kiksapa Consulting, LLC, Mandan, ND, USA

3 University of Arizona, Tucson, AZ, USA

4 Alaska Native Science Commission, Anchorage, AK, USA

5 Institute for Tribal Environmental Professionals, Flagstaff, AZ, USA

6 Intertribal Council on Utility Policy, Minneapolis, MN, USA

7 United States Geological Survey, Flagstaff, AZ, USA

8 Pacific Northwest Tribal Climate Change Network, Eugene, OR, USA

9 National Aeronautics and Space Administration, Greenbelt, MD, USA

10 National Wildlife Federation, Denver, CO, USA

discusses the synergy and discord between traditional knowledge systems and science and the emergence of cross-cutting issues and vulnerabilities for Indigenous peoples. The challenges of coalescing information about climate change and its impacts on Indigenous communities are outlined along with recommendations on the types of information to include in future assessment outputs. We recommend that future assessments – not only NCA, but other relevant local, regional, national, and international efforts aimed at the translation of climate information and assessments into meaningful actions – should support integration of Indigenous perspectives in a sustained way that builds respectful relationships and effectively engages Indigenous communities. Given the large number of tribes in the US and the current challenges and unique vulnerabilities of Indigenous communities, a special report focusing solely on climate change and Indigenous peoples is warranted.

"Climate change…remind[s] us that, as my Lakota relatives say, 'We are all related.' That might be the wisdom we need most whether scientist or non-scientist - Indigenous or non-Indigenous." – Dr. Daniel Wildcat, Haskell Indian Nations University[1]

1 Introduction

Indigenous peoples, lands, and resources in the US are currently experiencing a vast array of climate change impacts (Bennett et al. 2014; Houser et al. 2001; Larsen et al. 2014; Maynard 2002; 2014; McLean et al. 2009; Nakashima et al. 2012; Redsteer et al. 2013). Multiple social, environmental, economic, and political stressors increase the vulnerability of Indigenous peoples to climate change impacts, putting their livelihoods, communities, and deep connections to the natural and living environment at risk. Colonial-driven transformation of human and natural systems have pushed Indigenous communities to marginalized territories (Reo and Parker 2013) and forced many tribes onto restrictive reservations with limited options for food and safety (Lynn et al. 2013). This sets the stage for systemic impoverishment and injustice experienced by Indigenous peoples across the country and now is evidenced in the form of decreased adaptive capacity to climate change impacts (Maldonado et al. 2013b).

In the US, there are 567 federally recognized tribes, 34 state-recognized tribes, and many non-recognized Indigenous communities, including those within the US territories. Therefore, the chapter on Indigenous issues in the Third US National Climate Assessment (NCA3) only captured a small portion of the breadth and depth of climate change impacts and solutions occurring throughout the Indigenous communities in the US and its territories. Most of the authors of this paper were authors of the "Indigenous Peoples, Land and Resources" chapter in NCA 3 (Bennett et al. 2014; hereafter "Indigenous Peoples chapter"), as well as serving in significant other roles. Several authors of this paper who are Indigenous scientists working with Indigenous communities were given opportunities to take leading roles in the NCA process. For example, one author (Bennett) served as a member of the NCA's federal advisory committee that helped shape the NCA process overall. Two other authors (Hiza Redsteer and Chief) co-authored the Southwest Technical report on Indigenous people, which was the first time such a chapter was produced for the Southwest region's main technical input report. These authors found that although the Indigenous Peoples chapter represented the most comprehensive chapter on this topic in an

[1] This quote is from Dr. Wildcat in the Special Issue of *Climatic Change* (Maldonado et al. 2013a), which was initiated by the NCA's Indigenous Peoples chapter author team.

NCA synthesis report to date, the data available to quantify impacts to many Indigenous communities were insufficient and more documentation is needed.

This gap in knowledge leaves tribes (and rural communities) at a significant disadvantage and compromises the overall state of knowledge about climate change impacts. Nevertheless, the NCA3 process deliberately engaged Indigenous peoples and documented the climate change impacts they experienced. As a result of extensive discussion across author teams and within NCA's federal advisory committee, the challenges faced by Indigenous communities across all regions and sectors rose to one of the twelve key findings of the entire report: "Climate change poses particular threats to Indigenous Peoples' health, well-being, and ways of life" (Melillo et al. 2014:17).

Despite this progress, additional data are needed to comprehensively and sensitively represent Indigenous issues in future assessments. An ongoing, sustained assessment process (suggested by Buizer et al. 2013), as opposed to a business-as-usual quadrennial report summarizing the impacts of climate change, can more effectively identify and synthesize the relevant issues for the public and policy-makers. The process by which the Indigenous Peoples chapter was developed for NCA3 illustrates the depth of cultural engagement necessary for initiatives that effectively and adequately engage Indigenous peoples, but even this effort fell short in important ways.

Transforming the NCA into a more inclusive process and therefore creating a more comprehensive report requires: 1) meaningful inclusion and integration of Indigenous perspectives within other sectors of climate assessment reports; 2) focused chapters on Indigenous issues in all regional climate assessments; 3) more relationship-building, along with increased engagement and the inclusion of traditional knowledges in a culturally appropriate and Indigenous-led manner, and 4) publication of special reports on Indigenous issues, especially in circumstances where Indigenous issues have not historically been part of the climate assessment process. Additional efforts to support research with tribes on Indigenous lands are needed, to build capacity and provide technological tools that can strengthen statistical scientific confidence in observed impacts.

Recognition of the lack of meaningful engagement with Indigenous leaders, scholars, and resource managers in the first two NCA synthesis reports led to the significant effort to do a better job in developing the NCA3 Indigenous Peoples chapter. The first NCA report included a chapter on "Potential Consequences of Climate Variability and Change for Native Peoples and Homelands" (Houser et al. 2001). However, this chapter was based predominantly on discussions at a single 1998 NASA-sponsored Native Peoples-Native Homelands (NPNH) Climate Change Workshop in Albuquerque, New Mexico, and its subsequent report (Maynard 2002). For the NPNH workshop, nearly 200 Indigenous people from North America gathered together with climate scientists to examine the impacts of climate change and extreme weather variability on Native Peoples and their homelands from Indigenous cultural, spiritual, and scientific perspectives. Beyond this, the first NCA assessment leaders and chapter authors had limited engagement with and outreach to Indigenous peoples.

The second NCA report (Karl et al. 2009) did not include a chapter on Indigenous peoples. However, NASA sponsored a second NPNH Climate Change Workshop in Prior Lake, Minnesota in 2009. Although not specifically linked to the NCA process, the 2009 NPNH workshop was a collaborative effort of federal, regional, Indigenous, and non-Indigenous partners, including Haskell Indian Nations University, Honor the Earth (an Indigenous-led non-profit organization), NASA, and many other Indigenous partners. Over 400 participants discussed impacts of climate change on Indigenous peoples. Workshop attendees began developing response and adaptation actions and proactive recommendations to help ensure longer-term survival of Indigenous communities (Maynard 2014). A particularly significant outcome from the 2009 workshop was "The Mystic Lake Declaration" (Native Peoples Native

2009), demonstrating a growing strength of Native voices in the US. Members of the Indigenous Environmental Network presented the declaration to world leaders from more than 190 countries in Copenhagen, Denmark at the 15th Conference of the Parties to the United Nations Framework Convention on Climate Change. They contributed a powerful, unified Indigenous voice on the importance of Indigenous science and knowledge for addressing environmental and climate issues (Maynard 2014).

2 Developing the indigenous peoples, land, and resources chapter for NCA3

NCA3 staff and supporting agencies went beyond merely developing a summary of existing research, instead designing an inclusive process to develop and document the required foundational information. Activities began with the inclusion of Indigenous peoples in the preliminary workshops that framed the focus and organization of NCA3 (http://www.globalchange.gov/engage/process-products/NCA3/workshops). Additional process elements are described below.

2.1 Coalescing information about climate change and Indigenous communities across the US

The convening lead authors of the Indigenous Peoples chapter established a technical input team comprised of tribal members, agencies, academics, and non-governmental organizations to work together to identify relevant knowledge, data, and information to submit to the NCA as technical input, assess the appropriateness of referencing this material in the chapter, and identify gaps in the literature. The team also worked to ensure that input was submitted from sources throughout the US and represented a wide range of issues important to Indigenous peoples in the context of climate change. The Indigenous Peoples chapter incorporated technical input gathered from Indigenous communities, Indigenous scholars, and other researchers around the country that identified key climate change impacts to Indigenous culture, communities, natural resources, treaty rights, and sovereignty. Over 200 sources of technical input were submitted to the NCA, including published, peer-reviewed articles, workshop reports, Indigenous communities' activities and reports, and audio and video submissions of recorded interviews with Indigenous elders and community members.

Other chapters in the NCA also incorporated Indigenous-related technical input. Examples include the Great Plains chapter, which highlighted efforts of the Oglala Lakota Tribe in South Dakota to incorporate climate adaptation and mitigation within tribal sustainable development initiatives (Shafer et al. 2014). The Alaska chapter emphasized the increasing vulnerability of Alaska Native communities with a section specifically dedicated to research in Native communities that included traditional knowledges and the observations of elders and Alaska Natives in the region (Chapin et al. 2014). Additionally, the Coastal chapter highlighted the particular challenges of climate change and sea-level rise – as well as the unique barriers to adaptation – for coastal tribes (Moser et al. 2014). Despite these success stories, the inclusion of Indigenous knowledges and perspectives across other chapters was less evident.

Even with a chapter dedicated to Indigenous issues and some inclusion of Indigenous issues in other chapters, it quickly became clear that while a large volume of literature existed regarding climate impacts locally, regionally, and nationally, information specific to Indigenous peoples, their lands, and resources was seriously lacking. The climate literature covered regions encompassing Indigenous lands, but for the most part did not address Indigenous peoples' concerns or needs, nor did Indigenous peoples lead the development of the climate literature. To address this issue, over fifty authors representing Indigenous communities, academia, agencies, and organizations collaborated on a Special Issue for *Climatic Change*, the first peer-reviewed scientific journal issue devoted to climate-change-related impacts on Indigenous communities across the US (Maldonado et al. 2013a). This extensive effort, led by NCA staff, is evidence of the high priority that this topic was given in the NCA3 report development process.

During the development of the NCA3, several opportunities arose to conduct regional workshops to gather input and gain a more in-depth understanding of Indigenous concerns about climate change. These workshops resulted in reports from Coastal Louisiana tribes (2012), Great Lakes tribes (Shifting 2011), Alaska Native communities, Great Plains tribes (Riley et al. 2011), and the Pacific Islands (Souza and Tanimoto 2012). The workshops helped participants and the technical input team realize the need to consider and address not only the impacts of climate change on resources that many Indigenous communities depend upon and are deeply connected to, but to also acknowledge and highlight the grave risk climate change poses to entire cultures and ways of life. Fortunately, Indigenous communities themselves often hold the knowledge that can help them and others navigate towards a more sustainable path (see Fig. 1).

2.2 The selection of key messages

Many themes emerged from assessment findings; these included impacts on water resources, forest resources and ecosystems, evidence of multiple kinds of traditional knowledge, cultural practices, food subsistence, food security, adaptation, policy barriers, policy opportunities, displacement, relocation, risks specific to coastal communities, extreme weather events, human health, energy supply and use, and social and environmental justice. To condense the initial 49-page draft to the 3000-word-per chapter limit, the author team determined key themes by considering the most discussed issues in technical inputs, commonalities across regions, and the most immediate impacts and adaptation needs. Highlighted themes are: availability of and access to traditional foods; decrease in water quality and quantity; declining sea ice in Alaska; Alaska Native communities' exposure to health and livelihood hazards; and forced relocation.

Fig. 1 "The embers of our way of life must be carefully preserved by the ashes of time to light into flame the passion of our youth to assure that it endures." – Papalii Dr. Failautusi Avegalio. Photograph by Eren Avegalio

Two themes focus specifically on Alaska, where many Alaska Natives are experiencing the most dramatic and immediate effects of climate change, in part due to regional temperatures increasing at twice the global rate in the Arctic over the past several decades. Arctic Native peoples are observing and experiencing these changes on a daily basis (Cochran et al. 2013). Scientific research documenting environmental changes on Indigenous lands and waters, and impacts on Native livelihoods and cultures is limited, with the majority of studies documenting changes in Alaska and specifically the Arctic. In fact, the "Polar Regions" chapter of the 2014 Intergovernmental Panel on Climate Change (IPCC) Working Group II Report had three times more articles by or about Indigenous peoples in the Arctic than its 2007 report, significantly strengthening the inclusion of Native peoples in the Arctic (Larsen et al. 2014). While the NCA3's Indigenous Peoples chapter highlighted the themes that emerged most prominently through technical inputs, it could not capture all the effects of climate change occurring in and around Indigenous communities in the US due to the very stringent constraints on the length of the document.

3 Lessons learned, challenges, and recommendations for a sustained NCA

The primary challenge in writing the Indigenous Peoples chapter was that the majority of information on climate change impacts and adaptation and mitigation strategies for Indigenous peoples was not documented in peer-reviewed scientific literature. Many of the effects of a changing climate observed and experienced by these populations are talked about in oral tradition and traditional knowledge systems. These pathways for knowledge, however, are not often considered to have equal validity to scientific studies and findings. The revision and review process for the chapter highlighted the need to address discrepancies between inclusion of Indigenous peoples in assessments such as the NCA, and exclusion of Indigenous ways of knowing and observing from the current system for assessing climate change used by government agencies and Western scientists. This is a significant issue that should receive further consideration in future assessments.

Another challenge arose from tensions between how government agencies and scientists reviewing the Indigenous Peoples chapter explained and understood climate change and how climate change was described in technical inputs by Indigenous peoples. Some phrases and concepts do not translate easily between English and tribal languages. For example, in Navajo "climate change" is often expressed as a disruption of the balance and harmony of one's spiritual and cultural connection to Mother Earth and Father Sky. The authors worked to manage the divergent perspectives of Western science and traditional perspectives by explaining to reviewers that there are different ways of knowing and multiple kinds of traditional knowledge; these need to be given equal weight to Western forms of knowledge when discussing issues of importance to Indigenous people.

A third challenge – experienced across all NCA3 chapters – was that in the absence of dedicated funding, individuals had to volunteer their time as chapter authors or members of a technical input team, or find other sources of support. This limited who could contribute to the process and caused a tremendous strain on those who did participate. One of the most valuable lessons learned was that the current approach to assessments is not sustainable without additional resources. Although the technical input process was ultimately effective, the overall assessment process was unclear to participants and it was often overwhelming for the individuals striving to gather and synthesize information in a short amount of time.

Overall, increased Indigenous engagement is critical for more comprehensive NCAs in the future. As recommended in the special report on how to build a sustained assessment (Buizer et al.

2013:27), "The NCA should ensure adequate support for tribal engagement in future assessments, and include tribal engagement as a metric of success for ongoing efforts." For an assessment to be comprehensive and inclusive, Indigenous peoples must remain engaged. Additional information on barriers to adaptation, research, and capacity building in Indigenous communities should be a major focus in future assessments. Additional efforts are needed to incorporate contributions from traditional knowledge systems into assessments, because these data are primarily from regions that are least monitored and least studied (Redsteer et al. 2012). Huntington (2000) provides some guidance on inclusion of traditional knowledges that could serve as a point of departure.

Activities beyond the assessment itself are also needed to support adaptation efforts in Indigenous communities. In the absence of more concerted adaptation efforts, many threats to traditional life-ways may be left unaddressed or not given adequate emphasis. Efforts to expand research on Indigenous lands should include comparisons of outcomes from conventional scientific tools with traditional knowledge to bolster statistical confidence required by the assessment process. Very few data are available that quantify changes or establish baseline conditions for many Indigenous communities. Much is left out of assessment reports simply because local changes – which are sometimes clearly evident – have not been documented in scientific literature. The scant studies available indicate that some Indigenous communities are already experiencing significant impacts, but research in Indigenous communities has been uneven.

Expansion of Indigenous engagement on an ongoing basis would allow for a more comprehensive NCA, e.g., by enabling collection of additional data between assessments and by providing funding and capacity to carry out Indigenous-directed scientific studies on climate change impacts. Such effective engagement requires long-term relationships and investment. Important documents already developed by Indigenous peoples can be used to guide the establishment of a permanent and formalized structure for Indigenous participation in the NCA (see, for example, CTKW 2014; Grossman and Parker 2012; and Johnson et al. 2013).

On the basis of these lessons and our overall experience, and to address the challenges listed above, we make the following recommendations to enhance the inclusion of and engagement with Indigenous peoples in a sustained NCA and create a more sustainable process:

3.1 Include a dedicated indigenous peoples, land, and resources chapter in future NCA synthesis reports

To produce a more comprehensive NCA that expands knowledge over time, an Indigenous Peoples chapter must be included in every NCA synthesis report that is deliberately designed to fill gaps in earlier reports. Enhanced Indigenous inclusion in the NCA assessment process would support this effort.

To foster this engagement, at least one person representing Indigenous peoples should be invited to be on each NCA federal advisory committee charged with writing NCA synthesis reports. The committee should also include other members who are either Indigenous or who work with Indigenous peoples on issues related to climate change. Beyond the chapter on Indigenous peoples, Indigenous leaders and scholars should be included as authors on regional and sectoral chapters. This may lead to new understanding of climate change impacts and adaptation strategies as well as increased knowledge exchanges. Furthermore, future NCA publications should include Indigenous leaders and scholars as peer-reviewers, not just for Indigenous publications but also for all aspects of the climate assessment to ensure that Indigenous issues are addressed holistically and comprehensively.

3.2 Provide financial resources for Indigenous representatives to effectively participate in the NCA

As discussed in the special report on building a sustained assessment,

> "[Indigenous peoples] participation is critical to the success of a sustained assessment, by providing science support and data collection, including Indigenous knowledge, on Indigenous peoples' lands. However, the ability of Indigenous communities to participate in ongoing and emerging assessment related activities is seriously hampered by a lack of resources. The USGCRP agencies, as appropriate, could make funding available to Indigenous organizations for production of technical inputs and coordination of these inputs as part of a sustained assessment" (Buizer et al. 2013:27).

A number of groups already work with Indigenous peoples on climate change issues and represent a variety of viewpoints, regions, and sectors (see Table 2). The US Global Change Research Program (USGCRP) should work with these groups to gather information from their meetings and discussions and produce reports between the quadrennial synthesis reports that can be used as input for the Indigenous Peoples chapter. The Institute for Tribal Environmental Professionals (ITEP) at Northern Arizona University, for example, produces climate change profiles describing impacts and adaptation strategies being implemented by Indigenous nations across the US. Such entities could organize technical input meetings, produce technical input reports and special reports, and serve as the interface between NCA and Indigenous communities throughout the US and its territories.

One option for supporting ongoing assessments is to provide assistance to organizations that are already working on climate-related issues with Indigenous communities. Additionally, a liaison should be named to be part of the NCA process, to facilitate the connection between Indigenous communities and USGCRP and manage and coordinate a number of elements across the NCA process, such as those described below. A sustained assessment will be more effectively achieved by working with these established entities that are already adept at dealing with sovereign nations and state- and non-federally recognized Indigenous communities.

3.3 Provide networking support for indigenous groups

A team of Indigenous reviewers should monitor technical inputs for all sectoral and regional chapters, to help ensure that issues faced by Indigenous peoples are addressed in an integrated and holistic way rather than isolated and separated from governance structures and assessments, research agendas, and plans to address climate change. Moreover, inclusion of Indigenous input will facilitate the assessment of climate impacts across jurisdictional boundaries. Indigenous authors and reviewers are likely to remain engaged if they feel that they can help to shape the assessment process, that their voices are given equal and respectful consideration along with climate scientists, and that their participation is helping forge effective partnerships to support the actions needed for effective climate adaptation.

This networking effort could be better integrated and leveraged through NCA's existing network of partner organizations, NCAnet. The NCA should reach out to trusted Indigenous-led networks and organizations in each region to more effectively engage and invite tribal leaders and those holding leadership positions in a tribe, Indigenous communities, Indigenous networks, tribal colleges and universities, and Indigenous and non-Indigenous organizations that work on Indigenous-related climate change issues to become members of NCAnet.

To facilitate a wider array of Indigenous perspectives and increased Indigenous input to the NCA process, we recommend convening an annual meeting that brings together Indigenous networks and provides a technical input report to the NCA. Rising Voices, Indigenous People's Climate Change Working Group, and Shifting Seasons are organizations that provide successful models for event organization. Smaller workshops should also be held in different regions of the country and workshop reports produced to provide input for future assessments. A variety of Indigenous, federal, and state partners could be involved in convening such meetings. For example, the Great Plains intertribal workshop held in Oklahoma (Riley et al. 2011) was hosted by Haskell Indian Nations University (a tribal college) in collaboration with the Oklahoma Climatological Survey (a state partner) and the Southern Climate Impacts Planning Program (a federal partner). Workshop reports submitted to the NCA as technical inputs should also be included in the NCAnet toolkit, which includes materials produced by NCAnet partners, and is posted on the NCAnet website (http://ncanet.usgcrp.gov/partners/resources).

3.4 Produce an NCA special report on indigenous peoples, land, and resources

Because of the limited word count for each chapter, a significant amount of information gathered for NCA3's Indigenous Peoples chapter could not be included in the final report. How climate projections in regions directly relate to Indigenous communities and examples of implemented adaptation strategies mostly had to be omitted. Consequently, the chapter could not provide in-depth information about barriers to adaptation or research, policy, and information needs. An Indigenous Peoples, Land and Resources Special Report for the NCA could serve as a significant benchmark for the state of research, information, mitigation, and adaptation needs for these populations.

Information in such a Special Report should complement the primary NCA report's contents and provide additional information specific to Indigenous groups, including: environmental changes that Indigenous peoples are observing; impacts being experienced; barriers to adaptation that Indigenous peoples are encountering; additional stressors that increase Indigenous peoples' vulnerability to climate change and limit their adaptive capacity; adaptation strategies being implemented; support needed to protect cultures and traditions; information, research, and capacity-building needs; and how needs identified in the NCA3 report are being addressed.

The Special Report should include an in-depth focus on key messages addressed in NCA3's Indigenous Peoples chapter, as well as other themes relevant to Indigenous peoples, and include information from across regions. It should serve as a foundation document for future assessments and be revised on an annual basis to form an annual technical input report, similar to reports produced for regions and other sectors (e.g., Dalton et al. 2013; Garfin et al. 2013). Furthermore, the Special Report and related products, such as factsheets and an executive summary, would serve as significant documents to inform decision-makers at tribal, federal, state, and local government levels.

Based on the key messages from NCA3's Indigenous Peoples chapter and the proposed Special Report, a list of Indigenous-relevant research questions should be provided to agencies, research organizations, funders, and decision-makers (see Table 1). Indigenous-related networks could contact tribal colleges and other entities that have extensive outreach components (e.g., Bureau of Indian Affairs, National Congress of American Indians, ITEP, Intertribal Timber Council) for their research findings, research needs, and emerging issues, as well as send requests to federal science networks, such as USDA's regional climate hubs and DOI's climate science centers, to address the questions that arise from the Special Report that align with the scope of their work.

Table 1 Examples of potentially significant research questions based on the five key messages from NCA3's indigenous peoples, land, and resources chapter

Key message	Examples of research questions
Climate change impacts threatening traditional foods	• How are the impacts of climate change on traditional foods affecting food security and the health and nutrition of Indigenous peoples? • How are changes to the quantity and composition of traditional food species impacting tribal culture and traditional tribal access?
Decrease in water quality and quantity	• How will climate change and other vulnerability factors such as population growth and land use changes affect the quantity and quality of surface and groundwater on and around Indigenous lands? • How can Indigenous water supply infrastructure be better adapted to the climate changes occurring (e.g., drought, permafrost melting, algal blooms)? • What are the potential impacts of climate change on tribal water rights (both in terms of quantity and quality) and off-reservation rights to fish, hunt, and gather (Cozzetto et al. 2013)?
Declining sea ice in Alaska	• What technologies and resources are available for Arctic Indigenous communities to protect their homelands from storms and floods due to declining sea ice? How have these technologies worked in other communities? • What impacts will the expected ice-free Arctic Ocean have on Arctic Indigenous communities and how can communities engage in potential benefits and challenges? • How does declining sea ice affect species that are ice dependent and that Indigenous communities depend upon for subsistence (e.g., polar bears, walruses, and seals)?
Alaska Native communities exposed to health and livelihood hazards	• What human health impacts may result from conditions of climate change (e.g., flooding, wildfires, sewer/water contamination, ice dependency for subsistence, and rapidly changing weather conditions)? • How can Alaska Native water supply, wastewater, stormwater, transportation, and energy infrastructure be better adapted to the climate changes occurring (e.g., permafrost melting, warmer temperatures, algal blooms, potentially moving disease vectors) so as to protect the health of Alaska Natives?
Forced relocation of entire Indigenous communities	• What types of governmental processes (tribal and federal) can be put into place to assist Indigenous communities with identifying lands and funds for relocation if that becomes a necessity (Cozzetto et al. 2013)? • How can Indigenous communities continue to viably live in place and ensure the protection of cultural sovereignty while community-led relocation is planned? • What are examples of community-led relocations already taking place and the lessons that can be learned from these experiences?

3.5 Build capacity for adaptation

Compared to an isolated report, a sustained assessment requires more proactive capacity-building activities that strengthen skills and abilities at the individual, community, organization, and government levels. These include getting the backing of tribal leaders; improving the ability of tribal managers to conduct and report on vulnerability and risk assessments; developing effective partnerships to support climate adaptation efforts and Indigenous-specific research questions; working to break down barriers to adaptation; sharing lessons learned about climate change impacts; and identifying steps needed to effectively implement and review mitigation and adaptation actions.

Information gathered should focus on describing a changing relationship between communities and the surrounding lands and waters, including examples of positive, forward-thinking adaptation plans and actions, and lessons learned from actions implemented by Indigenous communities.[2]

Federal agencies should provide support for research, education, and training of Indigenous students, faculty, community groups, and others in Indigenous communities with respect to using assessment information for climate adaptation. These activities should extend to Bureau of Indian Affairs schools, reservation grant, private and public schools, tribal colleges and universities, community groups, non-profits, and tribal and Native leaders. Federal internship programs targeted for climate assessment in Indigenous communities should be increased.

The scientific and cultural competency of both Indigenous and non-Indigenous community members and scientists to work together more effectively will be enhanced through research that embraces the co-production of knowledge and includes partnerships involving Indigenous communities, federal agencies, state and local governments, research institutions and academics, and non-profit and non-governmental organizations (see Table 2).

USGCRP and other federal agencies could increase partnership opportunities between Indigenous individuals and the science and assessment communities (of federal and state agencies and academic institutions) by integrating issues of relevance to Indigenous communities (such as traditional knowledge about climate and land changes) into their requests for proposals. Indigenous partners have a deeper understanding of the challenges that face their communities and will be the ones to implement adaptation actions and solutions, and so will make valuable and essential contributions to these partnerships.

Subsequent assessments should be guided by a living document that both provides new emerging issues and information, as well as summarizes key capacity-building and other needs identified in the proposed Special Report on Indigenous Peoples, Land and Resources and through the technical input process, such as the suggestion in the Indigenous Peoples chapter of NCA3 that "New governance institutions, frameworks, and funding mechanisms are needed to specifically respond to the increasing necessity for climate change induced relocation" (Bennett et al. 2014:307). The living document could be continually produced through documents uploaded to a portal, which could be hosted by USGCRP, allowing for the ideas summarized in the Special Report and technical input process to continue to evolve and develop. The document could provide input to federal and regional agencies as they develop research plans and requests for proposals and serve

[2] These ideas were shared at the Indigenous Peoples Climate Change Working Group (IPCCWG) meeting on March 14–15, 2015 at the Southwest Indian Polytechnic Institute (SIPI), in Albuquerque, New Mexico.

Table 2 Suggestions for agencies and researchers on how to most effectively partner with Indigenous communities

Work through trusted and respected regional and sectoral Indigenous networks and organizations to connect to local Indigenous communities. Some such networks include the Institute for Tribal Environmental Professionals, Alaska Native Science Commission, Alaska Native Tribal Health Consortium, Pacific Northwest Tribal Climate Change Network, Intertribal Council on Utility Policy, Indigenous Peoples Climate Change Working Group, Rising Voices, Shifting Seasons, Indian Nations Conservation Alliance, and others.
Build partnerships between Indigenous and non-Indigenous agencies and researchers and maintain communication over the long-term, in a sustained relationship.
Ensure that Indigenous communities' governments or councils have knowledge of, and approve of research that directly impacts their community.
Research should be done *with*, not *on* or *for*, Indigenous communities.
Indigenous peoples should be co-investigators and lead the management of project objectives and resources.
Tribal informed consent forms, research guidelines, and a statement of purpose should be established and agreed to that include the research methods and objectives and how traditional knowledge will be protected in the research process, findings, and outcomes.
Allocate funding and technical resources for Indigenous communities' capacity building that increases the ability of Indigenous communities to directly engage in climate change research and assessments.

as a companion piece to another document containing relevant research questions, such as those noted in Table 1. At the annual meeting proposed above, participants could evaluate how well these needs are being addressed and submit their findings as technical input to the NCA.

NCA3 was released online to increase its accessibility to a wider audience. However, a solely online platform discounts those who lack Internet and computer access, which is the case for a comparatively high percentage of Indigenous people. To increase accessibility of climate information, federal science networks should inform Indigenous governments within their regions about the NCA through presentations or direct communication with tribal consortiums or individual tribal governments and leaders. Federal science networks should also distribute printed copies of the NCA Overview and Highlights documents (http://nca2014.globalchange.gov/downloads) to Indigenous communities, as well as videos that have been produced related to the NCA. Some of the videos should be translated into Indigenous languages to reach Indigenous peoples who do not speak English and who are often the ones that primarily carry traditional knowledges and pass sustainable practices to the next generation.

Furthermore, the globalchange.gov portal could include a means for communities to submit not only climate-related data as technical inputs, but also specific research and support needs. A well-publicized mailing address should also be provided for submissions from those who lack Internet access.

3.6 Honor and recognize the value of traditional knowledges

It is essential that the value of traditional knowledges be honored and recognized, and what is sacred to Indigenous communities and knowledge holders be protected. Future climate assessments and scientific studies should reference traditional knowledges as well as Western science, while respecting the right of tribes and Indigenous peoples to hold their traditional knowledges confidential.

The most recent IPCC Working Group II report concluded that traditional knowledge must be included when considering climate change adaptation (IPCC 2014). As stated by the Technical Summary,

> Indigenous, local, and traditional knowledge systems and practices, including indigenous peoples' holistic view of community and environment, are a major resource for adapting to climate change (*robust evidence, high agreement*). Natural resource dependent communities, including indigenous peoples, have a long history of adapting to highly variable and changing social and ecological conditions. But the salience of indigenous, local, and traditional knowledge will be challenged by climate change impacts. Such forms of knowledge have not been used consistently in existing adaptation efforts. Integrating such forms of knowledge with existing practices increases the effectiveness of adaptation. (Field et al. 2014:87)

Much of the knowledge about climate change impacts experienced by Indigenous peoples and their mitigation and adaptation strategies exist in oral tradition. A great deal of this information, particularly the observation of changes and how to adapt, is relevant for both Indigenous and non-Indigenous communities. Traditions other than Western scientific modes of expression need to be respected and considered as valid. Indigenous contributors to the NCA need to be able to tell stories from their communities in a culturally appropriate way that utilizes storytelling and does not derive solely from a Western, scientific perspective.

An ad hoc workgroup of tribal scholars, tribal leaders, and others developed a publication (CTKW 2014) in response to a call by the Department of Interior Advisory Committee on Climate Change and Natural Resource Sciences to increase understanding about the role of and protections for traditional knowledges in climate change initiatives. The publication suggests guidelines for considering the significance of traditional knowledges in relation to climate change and discusses the potential risks to Indigenous peoples in the US from sharing these knowledges in federal and other non-Indigenous climate change initiatives. This article deliberately uses the phrase *traditional knowledges* throughout the publication because *knowledges* are emergent from the symbiotic relationship of Indigenous peoples and places —a nature-culture nexus (CTKW 2014). The use of *knowledges* is intended to emphasize that there are diverse forms of traditional knowledge and knowledge systems that must be recognized as unique to each tribe and knowledge holder.

One of the greatest challenges in gathering technical input for the NCA was determining *what* "science" and *what* "knowledge" would meet the Data Quality Act requirements (see White House Office of Management and Budget 2002) to which all government-approved reports are held. More subtle, but no less difficult, was the challenge of getting experts trained in Western scientific procedures and standards to see the value and contribution of traditional non-scientific ways of knowing and honoring their importance. Technical input must include significant peer-reviewed literature. Yet much of the knowledge held by Indigenous peoples is traditional knowledge, uniquely Indigenous ways of understanding, defined differently by each tribe and each knowledge holder.

Sharing traditional knowledges in a national context is not always culturally appropriate and may put sacred knowledge at risk (CTKW 2014). And yet, traditional knowledges do inform Indigenous ways of understanding climate change impacts and adaptation solutions in a way that should be recognized and brought to bear on efforts such as the NCA.

4 Conclusion

While the NCA is a scientific document and does not make policy recommendations, it is a key document that policymakers turn to for formulating policy proposals related to climate change. Already facing the impacts of a changing climate, Indigenous leaders and communities should have their voices heard in this process, building upon their engagement and initiatives in international forums, such as the IPCC report highlighted above. Indigenous knowledges, ways of knowing, and worldviews must be respected. Incorporating Indigenous ingenuity, or "Indigenuity," and traditional knowledge systems in climate assessments and other efforts help to democratize these national and international processes (Wildcat 2009, 2013).

Climate change-related information shared by Indigenous communities in the US must continue to be not only included, but also expanded upon for the NCA to be truly comprehensive. This will require more work but will make the NCA a more compelling and complete document. Inclusion in the sustained NCA process could be tremendously beneficial in supporting Indigenous communities' needs regarding climate change.

A sustained NCA process would provide a key opportunity to bring together Indigenous leaders and practitioners, Indigenous elders, tribal college and university faculty and students, national and international Indigenous and non-Indigenous scientists and scholars, government agencies, and research organizations. This confluence will strengthen the voice of Indigenous peoples in the US on issues related to climate change and will contribute to the development of adaptation solutions at this critical moment.

References

Bennett TMB, Maynard NG, Cochran P, Gough R, Lynn K, Maldonado J, Voggesser G, Wotkyns S, Cozzetto K (2014) Ch. 12: Indigenous peoples, lands, and resources. In: Melillo JM, Richmond TC, Yohe GW (eds) Climate change impacts in the United States: the third national climate assessment. U.S. Global Change Research Program, Washington, DC, pp 297–317

Buizer JL, Fleming P, Hays SL, Dow K, Field CB, Gustafson D, Luers A, Moss RH (2013) Report on preparing the nation for change: building a sustained national climate assessment process. http://downloads.globalchange.gov/nca/NCADAC/NCADAC_Sustained_Assessment_Special_Report_Sept2013.pdf

Chapin FS III, Trainor SF, Cochran P, Huntington H, Markon C, McCammon M, McGuire AD, Serreze M (2014) Ch. 22: Alaska. In: Melillo JM, Richmond TC, Yohe GW (eds) Climate change impacts in the United States: the third national climate assessment. U.S. Global Change Research Program, Washington, DC, pp 514–536

Climate and Traditional Knowledges Workgroup (CTKW) (2014) Guidelines for considering traditional knowledges in climate change initiatives. https://climatetkw.wordpress.com

Coastal Louisiana Tribal Communities (2012) Stories of change: coastal Louisiana tribal communities' experiences of a transforming environment. Workshop report input into the national climate assessment. Pointe-aux-Chenes, Louisiana. http://data.globalchange.gov/report/coastal-louisiana-tribal-communities-stories-of-change-2012

Cochran P, Huntington OH, Pungowiyi C, Tom S, Chapin FS III, Huntington HP, Maynard NG, Trainor SF (2013) Indigenous frameworks for observing and responding to climate change in Alaska. Clim Chang 120(3):557–567

Cozzetto K, Chief K, Dittmer K, Brubaker M, Gough R, Souza K, Ettawageshik F, Wotkyns S, Opitz-Stapleton S, Duren S, Chavan P (2013) Climate change impacts on the water resources of American Indians and Alaska Natives in the U.S. Clim Change 120(3):569–584

Dalton MM, Mote P, Snover AK (2013) Climate change in the Northwest: implications for our landscapes, waters, and communities. Island Press, Washington, DC

Field CB, Barros VR, Dokken DJ et al. (2014) Technical summary. In: Climate change 2014: impacts, adaptation, and vulnerability. Part A: global and sectoral aspects. Contribution of Working Group II to the Fifth Assessment Report of the Intergovernmental Panel on Climate Change Field, CB, et al (Eds), Cambridge, United Kingdom and New York, NY: Cambridge University Press, pp. 35–94
Garfin G, Jardine A, Merideth R, Black M, LeRoy S (eds) (2013) Assessment of climate change in the southwest United States: a report prepared for the national climate assessment. Island Press, Washington, DC
Grossman Z, Parker A (2012) Asserting native resilience: Pacific Rim indigenous nations face the climate crisis. Oregon State University Press, Corvallis
Houser S, Teller V, MacCracken M, Gough R, Spears P (2001) Chapter 12: Potential consequences of climate variability and change for native peoples and homelands. In: Climate change impacts on the United States: the potential consequences of climate variability and change. National Assessment Synthesis Team, Cambridge University Press, UK. http://data.globalchange.gov/assets/e9/97/436129058f2107f4925aeec13ed8/nca-2000-foundation-report.pdf
Huntington HP (2000) Using traditional ecological knowledge in science: methods and applications. Ecol Appl 10(5):1270–1274
IPCC (2014) Fifth Assessment Report: Impacts, adaptation, and vulnerability. http://ipcc-wg2.gov/AR5/report/final-drafts/
Johnson JT, Louis RP, Kliskey A (2013) Weaving indigenous and sustainability sciences: diversifying our methods workshop. National Science Foundation
Karl TR, Melillo JM, Peterson TC, eds. (2009) Global climate change impacts in the United States. Cambridge University Press, UK. http://downloads.globalchange.gov/usimpacts/pdfs/climate-impacts-report.pdf
Larsen JN, Anisimov O, Constable A, Hallowed A, Maynard N, Prestrud P, Prowse T, Stone J (2014) Chapter 28: Polar Regions. In: Climate change 2014. Impacts, adaptation, and vulnerability. Contribution of working group II to the fourth assessment report of the intergovernmental panel on climate change. Cambridge University Press. UK
Lynn K, Daigle J, Hoffman J, Lake F, Michelle N, Ranco D, Viles C, Voggesser G, Williams P (2013) The impacts of climate change on tribal traditional foods. Clim Chang 120(3)
Maldonado JK, Shearer C, Bronen R, Peterson K, Lazrus H (2013a) The impact of climate change on tribal communities in the US: displacement, relocation, and human rights. Clim Change 120(3):601–614
Maldonado JK, Pandya R, Colombi B, eds. (2013a) Climate change and indigenous peoples in the United States: impacts, experiences and actions. Clim Chang 120(3). Reprinted 2014, Springer International Publishing, Cham, Switzerland
Maynard NG, ed. (2002) Native peoples-native homelands climate change workshop report. U.S. national assessment on climate Change. U.S. Global Change Research Program, NASA Goddard Space Flight Center
Maynard, NG, ed. (2014) Native peoples-native homelands climate change workshop ii final report: an indigenous response to climate change. November 18–21, 2009. Prior Lake, MN
McLean KG, Ramos-Castillo A, Gross T, Johnston S, Vierros M, Noa R (2009) Report of the indigenous peoples' global summit on climate change. April 20–24. United Nations University—Traditional Knowledge Initiative, Anchorage
Melillo JM, Richmond TC, and Yohe GW, eds. (2014) Climate change impacts in the United States: the third national climate assessment. U.S. Global Change Research Program, Washington, DC. http://nca2014.globalchange.gov
Moser SC, Davidson MA, Kirshen P, Mulvaney P, Murley JF, Neumann JE, Petes L, Reed D (2014) Ch. 25: coastal zone development and ecosystems. In: Melillo JM, Richmond TC, Yohe GW (eds) Climate change impacts in the United States: the third national climate assessment. U.S. Global Change Research Program, Washington, DC, pp 579–618
Nakashima DJ, McLean KG, Thulstrup HD, Ramos-Castillo A, Rubis JT (2012) Weathering uncertainty: traditional knowledge for climate change assessment and adaptation. UNESCO, Paris
Native Peoples Native Homelands (2009) The mystic lake declaration. Native Peoples Native Homelands Climate Change Workshop II: Indigenous Perspectives and Solutions. Prior Lake, Minnesota, November 21, 2009. http://www.ienearth.org/docs/TheMysticLakeDeclaration.pdf
White House Office of Management and Budget (2002) Guidelines for ensuring and maximizing the quality, objectivity, utility, and integrity of information disseminated by federal agencies. Fed Regist 67(36):8452–8460
Redsteer MH, Kelley KB, Francis H (2012) The observations of Navajo elders and the refining of our understanding of conventional scientific records: planet under pressure. International meeting on climate change, London
Redsteer MH, Bemis K, Chief K, Gautam M, Middleton BR, Tsosie R (2013) Chapter 17 unique challenges facing southwestern tribes. In: Garfin G, Jardine A, Merideth R, Black M, LeRoy S (eds) The assessment of climate change in the southwestern United States: a report prepared for the national climate assessment. Island Press, Washington, DC, pp 385–404

Reo NJ, Parker AK (2013) Re-thinking colonialism to prepare for the impacts of rapid environmental change. Clim Chang 120(3)
Riley R, Blanchard P, Peppler R, Bennett TMB, Wildcat D (2011) Oklahoma inter-tribal meeting on climate variability and change: meeting summary report. Norman, Oklahoma. http://www.southernclimate.org/publications/Oklahoma_Intertribal_Climate_Change_Meeting.pdf]
Shafer M, Ojima D, Antle JM, Kluck D, McPherson RA, Petersen S, Scanlon B, Sherman K (2014) Ch. 19: Great Plains. In: Melillo JM, Richmond TC, Yohe GW (eds) Climate change impacts in the United States: the third national climate assessment. U.S. Global Change Research Program, Washington, DC, pp 441–461
Shifting Seasons (2011) Shifting seasons Great Lakes tribal climate change summit. College of Menominee Nation Sustainable Development Institute. Keshena, WI. August 23–25, 2011. http://sustainabledevelopmentinstitute.org/wp-content/uploads/2013/01/CCSummitDraft2.pdf
Souza K, Tanimoto J (2012) PRiMO IKE Hui technical input for the national climate assessment, tribal chapter. PRiMO IKE Hui meeting, January 2012, Hawai'i. http://data.globalchange.gov/report/usgcrp-primo-2012
Wildcat DR (2009) Red alert!: saving the planet with indigenous knowledge. Fulcrum Publishing, Golden, CO
Wildcat DR (2013) Introduction: climate change and indigenous peoples of the USA. Clim Change 120(3):509–515

Climatic Change (2016) 135:127–141
DOI 10.1007/s10584-015-1512-1

The third national climate assessment's coastal chapter: the making of an integrated assessment

Susanne C. Moser[1,2] **· Margaret A. Davidson**[3]

Received: 27 February 2015 / Accepted: 16 September 2015 / Published online: 28 September 2015

Abstract Coastal areas are on the front lines of the impacts of climate change. The immediate impacts of temperature, precipitation and sea-level change affect rich but already threatened ecological systems and the most populated, highly developed, and economically vibrant regions of human activity on the planet. The specific vulnerabilities, impacts and adaptation options and activities vary greatly across the coastal areas of the US. The charge given to the coastal chapter team of the third US National Climate Assessment (NCA3, released in May 2014) was to discern the key vulnerabilities and most important cross-cutting concerns across the extensive coastline of the US. This paper is a reflection on what the coastal chapter team accomplished and how it was done (including author selection, staff support, technical inputs, the chapter development process, within- and cross-chapter integration, the review process, the delivery and high-impact release, the timeline of key assessment steps, and evaluation of the chapter development process). It concludes with eight lessons that might inform the activities of future collaborative author teams writing transdisciplinary, integrated assessment reports.

"This article is part of a special issue on "The National Climate Assessment: Innovations in Science and Engagement" edited by Katharine Jacobs, Susanne Moser, and James Buizer."

Electronic supplementary material The online version of this article (doi:10.1007/s10584-015-1512-1) contains supplementary material, which is available to authorized users.

✉ Susanne C. Moser
promundi@susannemoser.com

[1] Susanne Moser Research & Consulting, Santa Cruz, CA, USA

[2] Woods Institute for the Environment, Stanford University, Palo Alto, CA, USA

[3] Coastal Inundation and Resilience, NOAA Ocean Service, Charleston, SC, USA

1 Introduction

Coastal areas are on the front lines of the impacts of climate change. At the crosshairs of climatic changes occurring over land and those occurring over and in the oceans, coastal areas are increasingly experiencing the direct physical impacts of changes in temperature, precipitation and run-off, sea-level rise, coastal storms, and the changing physical, biochemical and ecological conditions of the oceans. While non-climatic pressures such as population growth, urbanization, pollution, conversion of land cover/use, and land subsidence are still often the dominant drivers of coastal change today, further sea-level rise, ocean acidification, and coastal temperature changes will combine with these existing pressures to create increasingly difficult challenges for coastal communities (Moser et al. 2014; Wong et al. 2014).

The charge given to the coastal chapter team of the third US National Climate Assessment (NCA3) was to discern – in a concise (8-page), multi-disciplinary, integrated assessment – the key vulnerabilities and most important cross-cutting concerns and adaptation activities across the extensive and varied coastline of the US, ranging from the tropical island shorelines of Hawaii and affiliated island states in the Pacific and the Caribbean, to the Arctic shores of Alaska; and from the coasts of the Pacific, Gulf of Mexico and the Atlantic, to the shorelines of the Great Lakes (Fig. 1).

This paper reflects on what the coastal chapter team accomplished and how it was done (Sections 2 and 3), as well as what larger lessons might be drawn to inform the activities of future collaborative teams developing transdisciplinary, integrated assessments in the US and elsewhere (Section 4). Several other contributions to this Special Issue on the NCA3 offer insights into the development of other NCA3 chapters (e.g., Maldonado et al. (indigenous lands and resources), Grimm et al. (ecosystems), Moss (decision support)), revealing how the task of conducting an integrated assessment was approached by different author teams. The purpose here is thus not to convey the superiority of one particular approach, but to point to key procedural choices that helped accomplish the challenging task of delineating and communicating policy-relevant findings for US coastal managers, policy-makers, and stakeholders. We highlight how these choices differ from those made in other coastal assessments with which we are familiar to inform future assessors wishing to build on the accomplishments of the NCA3.

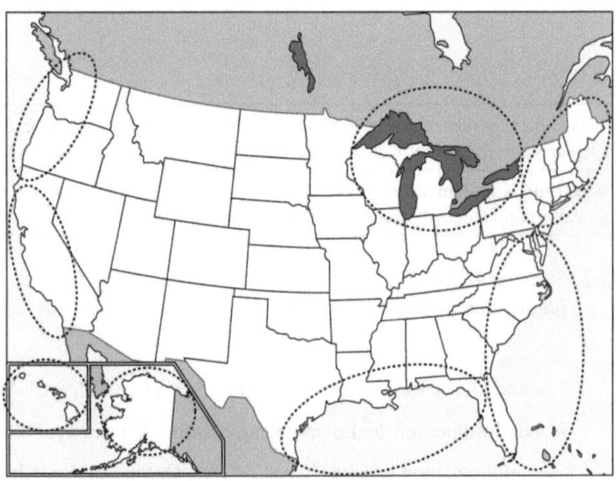

Fig. 1 US Coastal Regions Included in the NCA3 Coastal Chapter (Source: base map adapted from Wikimedia Commons)

2 Process and outputs: developing the coastal chapter's key messages

2.1 Author selection

A critical first step was the selection of the author team. NCA leaders had proposed the authors of this article as convening lead authors (CLAs) for the chapter, a choice partly motivated by our long-standing work in the coastal (adaptation) science and management arenas. Susanne Moser also served on the assessment's National Climate Assessment Development and Advisory Committee (NCADAC) and its Executive Secretariat. Margaret Davidson was then the director of NOAA's Coastal Services Center and co-author of a critically important underlying input document (Burkett and Davidson 2013). While the CLAs had known each other for about 20 years, we had not worked together on a project of this magnitude before.

The CLAs had discretion in proposing their team, though the final selection of authors required approval of the NCADAC because they had legal responsibility under their charter for the development of the NCA. Table 1 lists the formal (NCADAC) and informal (CLA-chosen) selection criteria. We considered the formal criteria minimal conditions that had to be met. Other critical considerations were whether the team would work well together, could deliver on time, and would be viewed as credible and legitimate by potential users of the assessment (i.e., we needed people with impeccable scientific *and* extensive experience in real-life coastal management). Thus, we weighed the informal criteria heavily, and they were critical to our collaborative success.

The resulting team (Paul Kirshen [University of New Hampshire], Peter Mulvaney [Skidmore, Owings & Merrill LLP], James F. Murley [South Florida Regional Planning Council], James E. Neumann [Industrial Economics, Inc.], Laura Petes [NOAA] and Denise Reed [The Water Institute of the Gulf], in addition to the CLA's) was evenly gender-balanced and included both seasoned and more junior experts. The team covered all necessary disciplinary expertise, and collectively had experience with all US coastlines. All team members had extensive experience working directly with coastal managers at local, state and federal levels – as managers themselves or as consultants and advisors – and all had had some experience working on other assessments before the NCA3 (e.g., local or regional assessments,

Table 1 Formal and informal criteria guiding author selection

Criteria guiding author selection	
Formal (given by NCA leadership)	• Professional expertise working on coastal climate change issues • Representation of US coasts and/or experience working in different coastal environments/region • Mix of disciplinary backgrounds (engineering, economics, geomorphology, ecology, governance/law, social science) • Balance of gender • Balance of senior/highly experienced and more junior individuals
Informal (CLAs' additional criteria)	• In touch with the reality of coastal managers/coastal adaptation challenges • Understanding of what an assessment is, how it is done • Ability to think interdisciplinarily and work transdisciplinarily • Availability and willingness to commit to intense volunteer work • Familiarity, respect, professional and personal connection

previous US National Assessments). This "real-world" experience was a critical advantage given the assignment to produce a chapter that was relevant to decision-makers. Several authors worked simultaneously on the IPCC Fifth Assessment. Despite the volunteer nature of the assignment, all of our first choice authors committed to several intense months (and ultimately two and a half years) of work.[1]

2.2 Staff support

Another crucial element of our process – and success – was highly competent staff support. Margaret Davidson detailed two staff members to assist the entire assessment development process. Both had previously supported the development of the Coastal Technical Input Report (Burkett and Davidson 2013) and thus were steeped in the material and process. Another chapter author, Laura Petes, had previously provided support to the Oceans Technical Input Report and was well integrated with the NCA coordination office team.

In addition to the NOAA staff, the NCA coordination office assigned staff to each chapter, including ours. Those served as chapter liaisons over the 2.5 years from beginning to end of the chapter development. Additional help came from two highly effective Chiefs of Staff at the NCA; two seasoned professional editors; and additional staff from NCA and NOAA's Technical Support Unit who helped with graphics. Because of our independent staff support, the coastal chapter team relied less on substantive help from the NCA coordination office than some of the other chapter teams. However, even with one CLA's direct knowledge of NCA3 guidance and internal procedures via membership on the NCADAC and its Executive Secretariat, the NCA staff support was important for the specific technical assistance provided, and for staying abreast of changing guidance to author teams, shifts in timelines, and overall coordination.

2.3 Technical inputs

One of the notable innovations in the NCA3 was its broad engagement of experts and stakeholders from outside the immediate assessment development apparatus (i.e., the NCADAC, NCA coordination office, author teams, and Federal agencies; see Buizer and Jacobs, and Cloyd et al., this issue). This engagement took many forms, but importantly involved a call for outside input in the form of "Technical Input Reports" (TIRs) and "assessment capacity" via Federal Register Notice (FRN 2011–17379; DOC 2011) and other communication vehicles. The FRN was released 6 July 2011, and inputs were requested by 1 March 2012 for full consideration by the author teams. This unprecedented request for data and reports from the public sent a signal that this NCA process was more open to new sources of information than previous processes.

Once the NCA coordination office had sorted through the material that was uploaded to the website – a total of 534 TIRs – the coastal team received those that were pertinent to its geographic and substantive focus. About 15% (or 79 discreet documents) were of coastal relevance. These documents included 17 sectoral and regional reports, including the coastal

[1] There is a significant difference in burden on authors depending on their employment status: for full-time, salaried individuals participation in an assessment effort does imply extra work and personal sacrifice while for soft-money researchers and consultants, participation means actual salary loss. Thus, authors must carefully assess their ability to commit to the required amount of work.

TIR previously mentioned that had been commissioned through federal agency activities; the remainder constituted input that may not otherwise have been discovered by, and included in, the assessment.

Of the 79 reports, the vast majority (89%) were scientific papers/reports/journal articles. The remaining nine documents included lists of references to other reports, news articles, factsheets, position statements, etc. The TIRs had been authored primarily by federal and state government employees, academics, technical consultants, but also by some NGO or private sector experts (e.g., EcoAdapt, Rand, Zurich Financial Services), and representatives of tribal groups.

The coastal team reviewed this input prior to the first in-person meeting. In the final assessment chapter, many of the documents were cited, and all are available via the NCA website along with other underlying documents and data (http://www.globalchange.gov/engage/process-products/NCA3).

2.4 Scoping meeting, ongoing team interactions, and the development of the key chapter findings

One month after receipt of the TIRs, the author team came together for one full-day meeting to develop the outline of the coastal chapter (with additional time for team members to get to know each other). This one-and-only in-person team meeting had five principal goals:

(1) To build a team (none of us knew all other members prior to that first meeting);
(2) To develop a clear understanding of the assessment task;
(3) To delineate a storyline for the coastal chapter, including the identification of "key vulnerabilities" that would be featured in this short national synthesis;
(4) To create a working understanding of the key guidance given to all NCA chapter teams (e.g., on risk framing, vulnerability, communication of confidence and uncertainties, traceable accounts of our decision process); and
(5) To develop a detailed chapter outline (including initial ideas of supporting graphic material), assign writing tasks, and agree on a timeline.

A cross-cutting goal for the meeting was to foster an all-hands-on-deck, collaborative spirit, a respectful, dialogic working style, and an interdisciplinary, practice-relevant conversation that was aimed at higher-order thinking and integration across disciplines and regional experiences, rather than domination by any one issue, discipline, coastal setting, or individual.

Through a series of round table discussions, CLA explanations of NCA guidance, and World Café-style engagement (Brown et al. 2005; The World Café 2005), the team developed working drafts of four out of its eventual five key messages. These were identified through an interactive discussion within the team that drew from the collective expertise and experience, new and important insights since the last National Climate Assessment (Karl et al. 2009), the coast-relevant TIRs, and other scientific literature.

The identification of key vulnerabilities was aided in important ways by an intuitive translation of the NCA guidance on risk framing (NCADAC 2012). It defined a "key vulnerability" as a potential impact that "could be of such high consequence for society that stakeholders consider them to be 'key vulnerabilities'" (p.1). Such a designation may result from the impact's magnitude, timing, persistence/irreversibility, limited potential for adaptation, distributional aspects, likelihood, or other attributes (ibid.).

For many authors, the technical guidance document was confusing, due in part to the various uses and definitions of risks, vulnerabilities, and impacts in different academic fields and practice. To convey its meaning and intention, the CLAs suggested an entry into the discussion through several questions: "what keeps you up at night?"; "what do you see as the Achilles' heel of coastal management in the face of climate change?"; or "what are the key challenges or outcomes that potentially threaten the fundamental structure and functioning of the coastal system?".

These questions helped to elicit chapter authors' key concerns, which were the starting points for deeper explorations of the three basic dimensions that determine a system's vulnerability (e.g., Cutter et al. 2003; Eakin and Luers 2006; Turner 2010). These key concerns and their underlying drivers were then mapped onto the formal definitions of vulnerabilities and risk (Fig. 2).

Figure 2 served as a boundary object that enabled experts from different disciplines and practical contexts to communicate with each other more easily than would have been possible otherwise. Each identified key concern was traced back through the key components of vulnerability: (1) *exposure* to potential climate-driven threats (e.g., changes and extremes in temperature, precipitation, storms, and sea level); (2) concurrent non-climatic stressors, which typically affect the *sensitivity* of a system; and (3) *adaptation* (including options, costs, capacity, constraints, and the state of implementation). The probability of the exposure component was estimated and expert judgment on trends in concurrent, non-climatic stressors and on the various aspects of adaptation all influenced the identification of key vulnerabilities.

After this initial identification of key vulnerabilities (focused on infrastructure and lifelines enabling coastal communities' functioning; differential social vulnerabilities; the coastal economy; and the conditions of coastal ecosystems), each was discussed in detail, bringing everyone's disciplinary and practical/management perspectives to bear.

Following the in-person meeting, the team met almost weekly by conference call to carefully assess the key issues identified. Various authors took the lead on writing about at least one of the key vulnerabilities, solidly anchoring them in the extant literature. Iterative team work on draft sections of the chapter resulted in the truly interdisciplinary treatment of

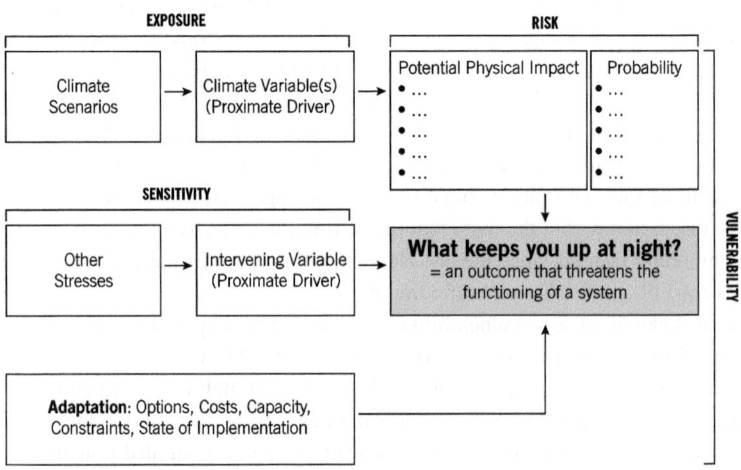

Fig. 2 Simplified Risk Framing Used in the Coastal Chapter (explanation in text) (Source: S. Moser)

each issue. In addition, regional examples and empirical data were integrated to flesh out each section, illustrating how these vulnerabilities were already emerging in US coastal communities today (not just in a distant future). Out of these discussions emerged the need to have a fifth key message, namely a cross-cutting one about the state of adaptation in coastal America.

Early consideration of the key messages and accompanying iconic graphics allowed us to draw on many additional outside contributors. For example, the Federal Highway Administration helped us develop a graphic showing coastal-inland transport of goods. In the case of social vulnerability in coastal areas, an EPA-led study (Martinich et al. 2012) had previously only covered selected US coastal areas. As social vulnerability rose to the top as one key finding of the chapter, an intensive effort was launched to update the underlying science and produce a consistent assessment of social vulnerabilities along all US coasts. Similarly, Thieler and colleagues at the US Geological Survey had completed a probabilistic assessment of shoreline change for some US coastlines. The prospect of being featured in the NCA3 encouraged them to finish their research for all US coasts (Gutierrez et al. 2014). Help also came from non-governmental sources: to provide a synopsis of climate change risks along with a range of adaptation responses from every coast, we approached academic experts and local, state, federal and tribal coastal managers in every region for examples, graphics, text, background material, and review to ensure accuracy. A considerable amount of work thus went into producing these graphics – often the most important elements for cursory readers.

These integrated products exemplify the benefits of truly interdisciplinary team work which advanced the state of knowledge far beyond that of any individual expert (see Jacobs and Buizer, this issue) towards enhanced understanding of both impacts and adaptive capacity.

2.5 Within- and cross-chapter integration and review process

Cross-chapter conversations were an essential aspect of ensuring consistency and accuracy, limiting redundancy, and integrating knowledge across sectors. Internal staff support was critical to this cross-chapter harmonization. NCA staff helped identify and reconcile repetitious or inconsistent material. In some instances, material that was better placed in one or another chapter was exchanged among chapters. Regional and sectoral expertise improved the robustness of chapter findings, and helped in the mutual reinforcement of carefully crafted messages.

Once the cross-chapter conversations were completed, the CLAs and internal support staff played a key integrative role in stringing the chapter sections together, harmonizing writing styles, and ensuring adequate backing from the literature. Multiple rounds of reviews further refined and improved the draft, including reviews from the Executive Secretariat, the NCADAC, the public, federal agencies, and a National Research Council (NRC) Review Panel. Once all of these comments were addressed (with every comment documented along with a response from the authors), a review editor evaluated the adequacy of the chapter team's responses. In several instances, this process resulted in additional revisions to ensure an adequate response. In some cases this required further collaboration with partners to fill key data gaps (e.g., from the National Flood Insurance Program). Once the chapter was nearing completion, additional detailed agency reviews, polishing with the help of the technical editors and, finally, a review from the White House (Legislative Referral Memorandum review),[2] adjudicated by the Executive Secretariat with the chapter authors, resulted in the final text (Fig. 3).

[2] This intense review process involving federal agencies and the White House is common procedure for all high-impact Federal Advisory Committee reports intended for adoption by the government.

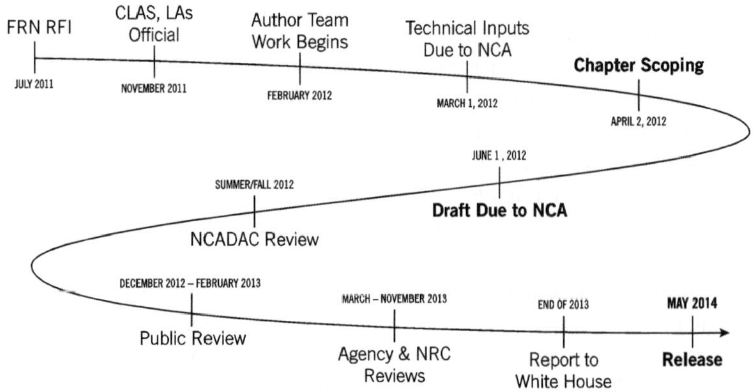

Fig. 3 Timeline of Chapter Development Notes: FRN RFI: Federal Register Notice Request for Information; CLAs: Convening Lead Authors; LAs: Lead Authors; NCA: National Climate Assessment; NCADAC: National Climate Assessment Development and Advisory Committee; NRC: National Research Council

2.6 Electronic delivery and high-impact release

On 6 May 2014 the Third National Climate Assessment report was approved by the NCADAC, accepted by the US government, and then released by the White House. In addition to White-House-generated media work, outside groups contributed in important ways to the high-visibility press coverage (Cloyd et al., this issue). Press conferences, a White House event with high-level officials, stakeholder calls, President Obama's interviews with TV meteorologists, ongoing press work for days following the release, and Congressional briefings were all part of the roll-out. One of the CLAs of the coastal chapter (SM) had the privilege of participating in the White House event with other NCA authors and administration officials (see: http://www.c-span.org/video/?319224-2/white-house-unveils-climate-assessment-report), and thus was able to bring attention to the coastal chapter.

The NCA3 was the first major assessment released electronically. Considerable work had gone into making this electronic version highly attractive, easily navigable, and readily accessible and linkable to other social media platforms (Facebook, Twitter). These innovations made the NCA3 publically accessible in unprecedented ways.

In addition, through an independently funded effort by The Story Group, findings of the assessment were also highlighted in freely available and widely advertised videos (see: www.vimeo.com/channels/nca). Some videos highlighted chapter findings and featured participating CLAs ("Scientists on the Frontlines of Climate Change"); others featured well-documented impacts experienced across the country, featuring individuals not involved in the assessments ("Americans on the Frontlines of Climate Change"). The coastal chapter video introduction was one of 10 videos released alongside the report in Mary 2014; eight additional ones have been made available since.

2.7 Chapter development timeline

The development of a first full draft of the chapter took approximately six months from the announcement of the chapter team, and a refined chapter ready for public review within one year (Fig. 3). While this may sound ample at first glance, a politically significant, fully integrated, 30-chapter national assessment requires multiple rounds of integration, fact checking, and internal review prior to release to external scrutiny. Thus, the actual preparation of the draft was accomplished in four months, with the remainder of that first year used for cross-chapter harmonization, internal review, and editing. Following the 90-day public review period and the review from federal agency experts and the NRC review panel, authors spent the second year in variably intense periods of responding to reviews, internal coordination, and harmonization. The final assessment development period involving White House review was largely delegated to the NCADAC's Executive Secretariat, but involved repeated checking in with chapter experts to ensure accuracy and full consensus on contents and wording.

Given the high policy-relevance of national assessments, there is little possibility for altering key, legally required steps in this process (e.g., public, NRC, agency and White House review). Staffing, author appointments, and larger procedural choices (such as requesting public input through an FRN) impose their own timing constraints. With the legal requirement of producing national assessments every four years, delays in author appointments, staffing, or process decisions inevitably affect the time remaining for the actual chapter development. Given the work involved, it is difficult to imagine a more compressed chapter and NCA development schedule unless NCA leaders were to resort to assessment processes that largely exclude public input and involvement. The only realistic way to give chapter authors more time for the actual content development is to start earlier after completing a previous assessment, or to sustain key elements of the assessment apparatus, so as to minimize ramp-up times (Buizer et al., this issue).

2.8 Evaluation

The NCADAC, NCA leadership and staff viewed evaluation of the assessment process and outcomes as a commitment to learning and improvement within the larger Sustained Assessment process (Buizer et al. 2013; USGCRP 2014). However, no independent, external evaluation has been launched to date. To capture immediate feedback before the author team dispersed, the CLAs of the coastal chapter developed a brief survey to request feedback from the lead authors as a way to capture the experience. While responses were not anonymous or shared with an independent evaluator, trust and respect within the team were high and authors were encouraged to speak frankly so that future assessments could be improved as necessary. Five of the six lead authors filled out the survey, the remaining responded informally. Web downloads and media coverage are also summarized below as a first indication of interest in coastal assessment findings.

2.9 Summary of unique traits of the coastal assessment

Key features make the NCA3 coastal assessment different from others we are familiar with (state, regional, previous US, and IPCC assessment efforts). They include the diverse composition of the author team involving academic and non-academic experts as well as coastal practitioners, and the rather unusual chapter leadership by a social scientist and a legal expert and policy-maker; a request for public submissions of TIRs at the outset of the assessment; the organization of the assessment around key vulnerabilities; the risk-based framing of these critical concerns; the transparent documentation of author choices in traceable accounts; and, through the iterative rounds of review, the elicitation of original research to fill key topical or regional gaps in understanding. Sophisticated electronic delivery and a related video, as well as intensive release and extended outreach efforts aimed to make the chapter (and NCA3 as a whole) more accessible than previous assessment products. Great emphasis was placed on team building, consensual framing, truly interdisciplinary assessment of the state of knowledge, and transparent decision-making regarding contents, wording, and levels of confidence on key findings.

3 Preliminary findings on coastal chapter reception and feedback on the development process

3.1 Chapter viewings, downloads, and media coverage

Due to the lack of an independent external evaluation of the coastal chapter's reception and utility in decision-making, little can be said to date about actual reach and uses of the document and related products. However, some data exist that are indicative of the interest in the chapter among assessment audiences. For example, between the release of the NCA (May 6, 2014) and January 28, 2015 (8.5 months), the full NCA3 report was downloaded 496,677 times from http://nca2014.globalchange.gov (data provided by USGCRP). In addition, the much shorter Highlights document was downloaded 47,746 times. Web analytics allow a more detailed breakdown: a total of 63,768 website visitor sessions with the 30 chapters were recorded. Of these, 28% were with the climate science chapter alone. Seven of the 10 most visited chapters were regional chapters (all of which contained a coastal focus); only two were sector-focused chapters (human health and agriculture). The full coastal chapter ranked 16th in web visitor sessions, 19th in individual chapter downloads, and 17th in web visitor sessions with the coastal segment of the report Highlights document.

With web traffic providing one indication of interest in the coastal chapter, media coverage of coastal assessment findings offer another. However, quantitative data on media coverage of the findings from the coastal chapter are not easy to provide. While NCA staff systematically track all assessment-related media coverage, numbers for citations per chapter are not available. Many news stories that quote NCA3 do not necessarily call out a particular chapter, so it is unclear whether reporters draw on the overview document, the climate science or coastal chapter or on findings from any one of the regions (all of which cover coastal concerns). However, a number of examples of news stories that call out coasts or sea levels is provided in Supplementary Material 1. According to the NCA's Engagement and Communications Coordinator, about 2,000 news stories were generated in the first week after the rollout (Cloyd et al., this issue), and "almost every story at the release used coasts as one of their primary touch points in reporting the effects of climate change on the US" (pers. comm.).

Finally, viewings of the videos produced to provide visual entries into the assessment could be tracked at the site where they were posted (Table 2). The comparative basis is limited to the 14 chapters and 4 additional climate impacts stories for which videos were produced. The videos were made freely available for download, so it is impossible to track how many additional times individual videos have been viewed from secondary sites (e.g., news sites, blogs, in classrooms). The numbers in Table 2 are thus only indicative of public interest in particular sections and findings of the assessment. Nevertheless, the coastal video is the most frequently viewed of the 14

Table 2 Viewing of video introductions to the NCA3

Sientists on the frontlines of climate change series		Americans on the frontlines of climate change series	
Chapter	Viewings	Chapter	Viewings
Coasts	3,563	Oyster farmer in WA	5,436
Hawaii & Pacific	3,040	Firefighter in CO	2,613
Southwest	2,392	Rancher in TX	1,899
Oceans	1,377	Farmer in IA	1,201
Alaska	1,351		
Agriculture	1,149		
Health	1.113		
Water	758		
Indigenous peoples	569		
Climate science	372		
Mitigation	263		
Rural communities	259		
Adaptation	216		
Southeast	153		

Data retrieved from www.vimeo.com/channels/nca on 7 July 2015

chapter videos to date and the second most viewed clip of all 18 (after one featuring oyster farmers in Washington State struggling with the impacts of ocean acidification on their business, an issue primarily delegated to the oceans chapter).

Future evaluations will have to illuminate how this interest in coastal chapter findings has translated into various uses in public policy, coastal management, and educational and outreach efforts.

3.2 Authors' evaluation of the chapter development process

A first-order internal evaluation was undertaken to capture immediate lessons about the deliberate procedural choices made and approaches used in developing the coastal chapter. This was thought to be important as feedback to the NCA3 leadership and to future assessment designers and leaders. Chapter authors were asked about their reasons for joining the assessment effort; composition of the author team; level of effort; the usefulness, frequency, and contents of the in-person meeting, teleconferences, and email updates; virtues and shortcomings of the final product; overall experience of the team collaboration.

Regarding the decision to join the assessment effort, authors wanted to be part of the process, "not sit on the sidelines," and "do something useful." Being part of a high-caliber author team, having an opportunity to learn, being part of a highly-visible, credible publication were also important draws. Notably, non-academic chapter authors appreciated the opportunity to contribute their practical experience and perspective, and to learn from and get closer to the climate science.

Generally, authors thought that good disciplinary breadth had been achieved, but there was less confidence that regions were treated evenly. While the voices of practitioners were influential in discussions, the writing burden remained largely on the more research-oriented team members. A suggestion was made to include someone from the development community on future assessment teams.

Authors felt the amount of work asked of them was more or less as expected and appropriate despite the intensity of demands on their time during certain periods. While contributors had relatively narrowly defined assignments, the framing and integrative work of the CLAs was essential for guidance and flow.

The in-person meeting was seen as essential for team building and for shaping the overall frame for the chapter. Several authors would have appreciated either a longer or a second meeting to deepen the joint work (but no resources were available for a second meeting). Due to time constraints, one felt the identification of key messages was too rushed, while others appreciated the early focus and development of a blueprint that guided all subsequent work. Similarly at odds were comments about the team conference calls and email updates: some viewed them as too frequent, others appropriate; some would have wanted more time for contents discussion, while others viewed them as necessary and helpful for exchange and staying abreast of NCA developments. While email updates on NCA procedures, developments, and chapter work were welcomed by some, others felt they did not sufficiently bridge the distance between the central NCA operation and those actually writing the report. There appeared to be no simple formula of how to strike the right balance between sheltering author teams from ongoing internal debates and shifts in procedure and yet keeping them informed of essential developments.

Overall, chapter authors were proud of the work they produced and felt the process led them as efficiently and effectively as possible to the final product. To them, the NCA3 assessment process was a worthwhile endeavor and was judged as more effective than participating in the IPCC or other assessments. While some wished the chapter had made even bolder statements, all were comfortable with the take-home messages. Some suggested a greater focus on pragmatic adaptation options would have improved the chapter.

Everyone noted that they learned from the process, and appreciated the leadership, procedural choices, and staff support. Most had not worked with each other prior to this assignment, and the NCA3 encounter resulted in several collaborative projects since. The shifting guidance throughout the chapter development, the amount of unpaid work, and the lack of clarity about how to produce the traceable accounts created frustration, pointing to the need for efficient learning so as to minimize these problems in future assessments.

4 Key lessons learned and conclusions

The coastal chapter CLAs brought a range of perspectives and experiences to this assessment, having previously studied, observed, formally evaluated, and/or served in various capacities in other assessments (as contributing or convening lead authors, reviewers, and review editors). Together with our expertise, these insights shaped our leadership style and vision for the process and product. While individual procedural choices and approaches we used may not be unique, in our experience they are uncommon. More important, however, is the question of whether "they worked." Based on the feedback from our author team, and reflections between us and our staff, we distill eight lessons that transcend the unique needs and challenges of a US national assessment and thus may have broader applicability.

Lesson 1 Competent, diverse assessment teams should include "boundary" and practitioner experts.

An assessment that covers diverse regional issues, involves different sets of expertise, should be relevant to decision-makers and be solutions-oriented, and that must be delivered in a short period of time requires a multi-skilled team. The criteria for selecting the author team, particularly the informal ones (Table 1), proved singularly important for shaping the "right" team for the task at hand. The critical innovation in this assessment was the inclusion of science-policy "boundary" workers and practitioner experts. Equally important was the time and resources spent in bringing authors together and fostering a collaborative, interdisciplinary spirit as a basis for the joint work. The positive reflections on the team work, the emerging collaborations beyond the assessment, and the willingness to do it again speak volumes of the importance of "the right people" involved and the emphasis placed on process and team-building. Thus, while products and processes have long been recognized as essential for effective assessments, choosing the right people may be the ultimate secret to success.

Lesson 2 Chapter leadership must have a direct link to the overall assessment leadership.

A complex, large assessment with numerous individuals in influential positions, and hundreds of others contributing their insights and ideas – even if lead competently and diligently – requires clear guidance, constant adjustments, careful attention, and meticulous coordination. It served the coastal team in crucial ways to have one CLA be close to these internal processes and to communicate relevant issues, translate guidance, and keep the author team abreast of developments. Anecdotally, other NCA3 chapter teams that did not have this direct linkage to the NCADAC reported greater confusion and disconnect, and less buy-in to the overall assessment process.

Lesson 3 A compelling consensual framing early on enables an easier assessment process.

Clear, consensual framing of the key issues and overall story that the assessment tells is essential for guiding the subsequent work, to keep inputs focused, and to identify necessary graphic and analytical input in a timely manner. Clearly, such an early framing should not become constraining and thus must remain open to modification. But together with boundary objects and common guidance, it proved to be a time-saving device and signaled to authors where and how to contribute efficiently and effectively.

Lesson 4 Awareness of audience(s), purpose, and context help deliver a more accessible, compelling and focused assessment.

The coastal chapter was one of many in a national assessment that responded to a federal mandate, and thus had to speak to key Congressional audiences as well as a broader national constituency. Keeping the purpose, audiences, and context (of 29 other chapters, the prevalent public conversation, and decision-making contexts of coastal managers) in mind proved crucial in the choice of key vulnerabilities, in importing or exporting certain text passages from or to other chapters, and in choosing language that is at once accurate, accessible, and compelling. Having communication expertise and end users on the chapter team was a crucial innovation in the NCA3 coastal chapter to help with these choices.

Lesson 5 Openness to and staff support for cross-chapter collaboration are a must.

The author team embraced the opportunity to collaborate not just with each other, but with other chapters, and with outsiders who could provide key information and data. This collaboration required a particular mind-set, commitment, time, and transparency. Crucially, the staff support available during NCA3 within the chapter itself and from the NCA central office was invaluable in enabling effective cross-chapter coordination. These ingredients, we believe, resulted in a richer story, more robust findings, greater engagement from outsiders, and ultimately increased receptivity from the audience.

Lesson 6 Commitment to a chapter vision is equally important.

The willingness to collaborate, compromise where necessary, and stay open to others' views, needs, and comments must be balanced with a willingness to defend one's vision, knowledge, and team. The chapter was refined over the course of at least 12 rounds of within-team, across-chapter, NCA-internal, and external review, making it more complete, more reliable, and more readable. But chapter authors and CLAs also defended the overall storyline that had emerged from a dialogue between academic and practitioner experts with their relevant disciplinary and management expertise.

Lesson 7 Given needs and constraints, the success of assessments hinges on people.

Most assessments are highly complex and largely unpaid efforts, conducted under inevitable time pressures, within unavoidable institutional constraints, and with an unrelenting demand to deliver a product that scores high – simultaneously – on scientific credibility, practical saliency, and procedural legitimacy. These constraints leave little room for changing the sequence and timing or for skipping key elements involved in conducting an assessment. To deliver successfully requires strong leadership, openness to learning, and a deep reservoir of collegiality, creativity, and team spirit, and fostering it in team meetings, calls and other direct interactions should be a top priority for all assessment leaders.

Lesson 8 External evaluation is essential to support the sustained assessment.

Whether or not the innovative approaches used in developing the NCA3 coastal chapter not just enabled a better assessment experience, but resulted in a more user-friendly, accessible, and influential product could not be ascertained by the author team. An independent, external evaluation is needed to answer the ultimate question of all assessments: did it make a difference in the world of policy-making and practice? Such an evaluation should generate comparative insights from multiple chapters (or other assessments) to ascertain how transferable these lessons are. Anchored as these are in the NCA3, as well as in our work on other local to international assessments, however, we believe they hold merit as "hypotheses" to be tested in future assessments. Only when evaluation becomes viewed as a useful learning aid (rather than as an effort in grading performance), however, can it provide continual strategic support to the sustained assessment process.

Acknowledgments The authors wish to thank the NCA3 coastal chapter team, its many contributors, the NCA staff family, and particularly Emily Wallace and Murielle Gamache-Morris for their assistance throughout the coastal chapter development. The latter also helped with the production of figures for this paper. Kathy Jacobs, Laura Petes, Emily Wallace, and three anonymous reviewers provided comments on an earlier draft of this paper, though none is responsible for the final outcome.

References

Brown J, Isaacs D, World Café Community (2005) The world café: shaping our futures through conversations that matter. Berrett-Koehler, San Francisco

Buizer J, Fleming P, Hays SL, Dow K, Field CB et al (2013) Preparing the nation for change: building a sustained national climate assessment process. Special Report National Climate Assessment and Development Advisory Committee, Washington

Burkett V, Davidson MA (eds) (2013) Coastal impacts, adaptation, and vulnerabilities: a technical input to the 2013 national climate assessment. Island Press, Washington

Cutter SL, Boruff BJ, Shirley WL (2003) Social vulnerability to environmental hazards. Soc Sci Quart 84:242–261

Department of Commerce (DOC), National Oceanic and Atmospheric Administration (2011) Technical Inputs and Assessment Capacity on Topics Related to 2013 US National Climate Assessment [Docket No. #110614333–1333–01]. Federal Register 76(134): 41217–41219, Notice on 13 July 2011, Washington

Eakin H, Luers A (2006) Assessing the vulnerability of social-environmental systems. Annu Rev Env Resour 31: 365–394

Gutierrez BT, Plant NG, Pendleton EA, Thieler ER (2014) Using a Bayesian network to predict shoreline change vulnerability to sea-level rise for the coasts of the United States. US Geological Survey Open File Report, USGS

Karl TR, Melillo JM, Peterson TC (eds) (2009) Global climate change impacts in the United States. Cambridge University Press, New York

Martinich J, Neumann J, Ludwig L, Jantarasami L (2012) Risks of sea level rise to disadvantaged communities in the United States. Mitig Adapt Strateg Glob Change 18(2):169–185

Moser SC, Davidson MA, Kirshen P, Mulvaney P, Murley JF, et al. (2014) Coastal zone development and ecosystems. In: Climate Change Impacts in the United States: The Third US National Climate Assessment, Melillo J, Yohe G and Richmond TC (Eds.), 579 –618, Washington: USGCRP

National Climate Assessment Development and Advisory Committee (NCADAC) (2012) Technical Guidance for NCA Authors: Risk Framing for Key Vulnerabilities Assessing Confidence Levels and Uncertainties. (Ver. 2012.02.10), Washington: NCADAC

The World Café (2005) Café to go: A quick reference guide for putting conversations to work. http://www.theworldcafe.com. Accessed 5 January 2015

Turner BL II (2010) Vulnerability and resilience: coalescing or paralleling approaches for sustainability science? Global Environ Chang 20:570–576

United States Global Change Research Program (USGCRP) (2014) Frameworks for Evaluating the National Climate Assessment: Workshop Report. Washington: USGCRP, http://www.globalchange.gov/engage/process-products/NCA3/workshops

Wong PP, Losada IJ, Gattuso J-P, Hinkel J, Khattabi A et al (2014) Coastal systems and low-lying areas. In: Field CB, Barros VR, Dokken DJ, Mach KJ, Mastrandrea MD (eds) Climate change 2014: impacts, adaptation, and vulnerability. part a: global and sectoral aspects. contribution of working group ii to the fifth assessment report of the Intergovernmental Panel on Climate Change. Cambridge University Press, New York, pp 361–409

Climatic Change (2016) 135:143–155
DOI 10.1007/s10584-015-1549-1

Assessing decision support systems and levels of confidence to narrow the climate information "usability gap"

Richard H. Moss [1]

Received: 5 May 2015 / Accepted: 23 October 2015 / Published online: 16 November 2015

Abstract This article focuses on the implications for the US National Climate Assessment (NCA) of diversifying information needs to support climate change risk management. It describes how the Third US National Climate Assessment (NCA3) evolved to begin to narrow the gap between information from climate and impact scientists and "intermediaries" (individuals who have expertise in climate science, communication, and decision-support processes)—who are sometimes collectively described as "producers" in this article—and the decision-making needs of a wide range of "users" (individuals involved in advising or making a wide range of policy and management decisions). One step in the evolution of the NCA3 included adding a chapter to assess decision-support tools and systems being used in climate-related decisions. Another involved efforts to improve characterization of the level of confidence of NCA3 authors in their findings to help decision-makers and their advisors differentiate well-established and more preliminary conclusions. This paper lays out an argument for increasing the role of the NCA in assessing decision-support systems in the Fourth Assessment (NCA4) and the Sustained Assessment. It also briefly reviews approaches and potential next steps related to characterizing uncertainty and communicating confidence intended to improve application of assessment findings by decision-makers.

The Joint Global Change Research Institute is part of the Pacific Northwest National Laboratory, which is operated for DOE by Battelle Memorial Institute under contract DE-AC05-76RL01830, and the University of Maryland. The author gratefully acknowledges the University of Arizona, Center for Climate Adaptation Science and Solutions, for providing intellectual and financial support during a period of residence as a visiting scholar when the first draft of this article was prepared.

This article is part of a special issue on "The National Climate Assessment: Innovations in Science and Engagement" edited by Katharine Jacobs, Susanne Moser, and James Buizer.

✉ Richard H. Moss
rhm@pnnl.gov

[1] Joint Global Change Research Institute, Pacific NW National Laboratory and University of Maryland, College Park, MD, USA

1 Diversifying information needs

The report produced by the Third US National Climate Assessment (NCA3) documents the increasing number of impacts being experienced across sectors and regions of the USA and projects that the pace of change and consequences for the nation will escalate (Melillo et al. 2014). "Billion dollar storms," and increased frequency of smaller "nuisance" events, are already motivating adaptation decisions (Bierbaum et al. 2014) and stimulating measures to limit emissions (Jacoby et al. 2014). As a result, an increasing number of people need information that enables them to understand the potential implications of climate change for their businesses, livelihoods, property, communities, and the natural resources on which all of these depend. Without such information, they are likely to make plans and investments that result in systems that are not fit for purpose and leave them vulnerable to climate change, thus increasing potential damages and threats to the health, safety, and prosperity of US citizens.

The report's sectoral and regional chapters illustrate the many routine management, planning, and investment decisions that are being affected by climate change. Table 1 lists a sample of major types of decisions that are being made in several sectors covered in the NCA3 report.

Table 1 Examples of decisions that could be affected by global change and benefit from access to existing imperfect knowledge

Sector	Decisions
Water resources	• Planning reservoir capacity • Managing water levels in the Great Lakes • Improving drought preparedness
Energy	• Planning peak demand loads • Siting bioenergy plantations • Expanding conventional energy production
Transportation	• Engineering fixed-route infrastructure • Increasing resilience of airports vulnerable to storm surge • Managing flood risk to subway systems
Buildings and related infrastructure	• Setting standards and design loads for structural safety (e.g., snow loads for roofs) • Sizing drainage systems and culverts • Siting buildings and related infrastructure relative to flooding, wildfire, or other hazards
Agriculture	• Redesigning food processing and supply chains for improved water and energy use efficiency • Investing in agricultural technology development and diffusion (crop varieties, post-harvest storage) • Anticipating needs for food security early warning and disaster management
Environmental conservation	• Using coastal ecosystems sustainably to support tourism, conservation, and fisheries • Long-term planning and budgeting to manage wildfires • Conservation planning to protect viable refuges with high potential to preserve biodiversity
Human health	• Establishing monitoring systems to track reemergence of familiar disease threats or emergence of new ones • Determining what public health investments are needed to manage heat stress in indigent populations • Monitoring changes in phenology to reduce exposure of sensitive populations to increased allergens

Looking across these decisions, they may be grouped into three broad categories. *Public policy* includes decisions about laws, codes, taxes, or other public mandates that set the overall regulatory framework in which public and private sector decisions are made. *Asset management* decisions take place in many sectors and are related to administering resources, infrastructure, or response mechanisms such as those intended to promote public health and limit damages from extreme events. *Infrastructure and resource planning* about infrastructure and natural resources concerns investments in or long-term plans for technologies, natural systems, or communities. While overlapping, each of these classes of decisions has particular characteristics including who is involved, duration of impact, the process used, and the type of information required. This has implications for the NCA and requires an increasingly diverse set of approaches to decision support.

These classes of decisions will require different types of decision-support systems to reflect the varied content of emerging issues, the complexity of decision-making contexts, and the multiple economic and societal interests affected. Lemos et al. (2012) identify a "usability gap" between current approaches to disseminating information from climate science and the needs of decision-makers. They observe that "narrowing the climate information usability gap" and "produc[ing] usable climate information to meet societal risk and adaptive management needs" requires that "we must rethink the ways in which we design and promote use-inspired basic and applied research programmes…." Clearly, new approaches to assessment and decision support are needed.

Perhaps the most fundamental change to the NCA process that will facilitate closing the usability gap is the adoption of a more fluid and interactive *sustained assessment* process that would generate a broader set of products to support the increasingly diverse information needs identified (see Buizer et al. 2015). A special report to the US government ("Preparing the Nation for Change: Building a Sustained National Climate Assessment Process," Buizer et al. 2013) described and recommended an evolving framework for connecting scientists and practitioners from the government, the civil society, the private sector, the tribal communities, and other parts of society. The process envisioned decentralizing assessment by moving beyond reports produced by the federal government to one in which localities and organizations contribute to a broader range of assessed products to use in conducting their own appraisals of the implications of climate change for their own interests. Some of these products may be produced and disseminated from federal agencies, but others may arise from the "bottom-up," for example, when an non-governmental organization (NGO), university, or consultancy adapts knowledge for a particular user in a way that can be generalized and applied by others with similar problems or circumstances. The sustained assessment shifts from *production* and *transmission* of scientific information through reports to *coproduction* of scientific information with a much greater level of interaction among decision-makers and producers of information. Transition to the Sustained Assessment will both create an increasing supply of decision-support products as well as the demand for more rigorous evaluation and assessment of these resources.

2 Evolution of decision-support tools and systems

Decision support refers to organized efforts to facilitate the use of knowledge to improve decision outcomes (Moss et al. 2014). In the context of this article, I will differentiate between decision-support *tools* and decision-support *systems*, noting that there is no broad consensus on these definitions in the literature. Decision-support *tools* include databases, simulation models, maps, reports, and other science-based products that provide information. Decision-support *systems* are knowledge frameworks that structure decision support processes (interactions of producers and users of climate and impact science); frame information needs; access and organize information; integrate information; inform judgments and preferences about options and tradeoffs; and sustain communications. Decision-support systems add value to tools (the information products) by providing interfaces or other approaches that assist the people or groups involved to apprehend and use the information content of decision-support tools in decision-making.

Decision support systems are designed to help overcome cognitive, communication, or other barriers and build shared understanding of the implications of scientific information for decisions. They can help structure stakeholder engagement and improve synthesis of information through approaches such as deliberative-modeling, cost-benefit analysis, and risk assessment. The distinction between *systems* and *tools* calls attention to the importance of evaluating both the information content and the approach used to facilitate the interaction of individuals or groups with that content in the context of their decision-making. Usability of the information provided by a decision-support system depends not only on the soundness of the basic science for the intended use but also on the quality and effectiveness of the approaches for presenting information and assisting participants to apprehend and work with it. A decision-support system must specifically account for the different capacities, preoccupations, perceptions, and needs of its intended users, as well as the characteristics of the institutions in which they are working.

Decision-support systems can provide assistance at any of several stages of a decision process. Figure 1 depicts a quasi-circular idealized adaptive decision-making process that includes the major steps of (1) framing the decision and information needs, (2) discovering and coproducing information, (3) integrating values, science, and other contextual factors, (4) deciding and implementing, and (5) monitoring, learning, and reviewing decisions and decision support (Moss et al. 2014). Decision-support systems can help to structure stakeholder engagement in decision-making processes, clarify information needs, access and organize information, integrate and analyze relevant factors to weigh tradeoffs or synergies, or structure data collection and analysis to support monitoring and adjustment. Systems can draw on knowledge and insights from a wide range of disciplines including engineering, business,

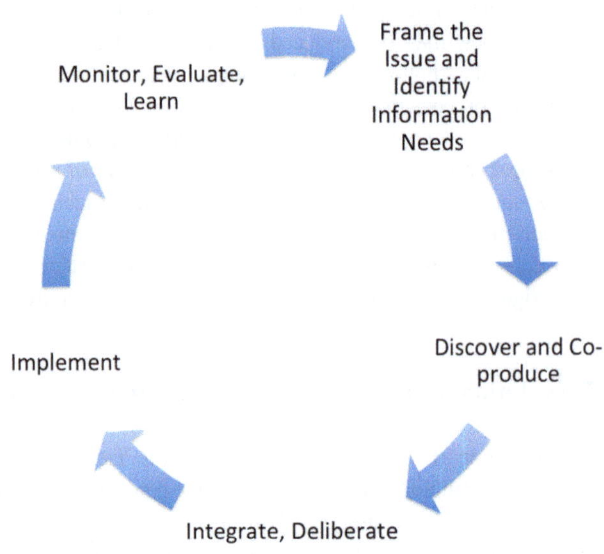

Fig. 1 Idealized adaptive management decision-making process. Decision-support systems and tools can help participants confront challenges associated with each of these stages of decision-making

economics, statistics, psychology and other social and applied sciences, climate science, hydrology, ecology, health sciences, geography, remote sensing, and many others. Decision-support systems can be used to address a number of barriers to implementing adaptation and mitigation (for adaptation barriers, see Eisenack et al. 2014; Ekstrom and Moser 2014; Moser and Ekstrom 2010; for mitigation barriers, see Bazerman 2009; Gifford 2011). They help establish conditions associated with successful consideration of climate change in decision-making (United States Government Accountability Office 2013).

Traditional climate change assessment reports such as those of the Intergovernmental Panel on Climate Change (IPCC) and previous NCAs are a specific type of decision-support product and have addressed a wide range of topics about basic climate science, impacts, and response strategies. They have informed decision-making, particularly for public policy issues, for example, related to the objectives and provisions of the UN Framework Convention on Climate Change. These reports survey and synthesize research from many disciplines, communicate the state of science related to identified issues, and identify gaps and research opportunities. They will remain important means of conveying information for policy- and decision-making.

But in response to the diversifying needs described above, decision support for climate change is evolving into a much broader set of approaches. Climate change decision-support tools and systems can, for example, map climate stressors and their impacts; portray expert judgments about changes in impact-relevant climate variables; estimate past and potential future carbon storage; provide data on land cover and use classifications in vulnerable areas such as coastal zones; provide geospatial analysis of floods, droughts, heat waves, and other stressors onto infrastructure or population groups; analyze extreme-event return periods; estimate energy demand, peak loads, and costs under different climate scenarios; provide indicators to inform managers about changes in climate exposures, species distributions, and natural resources; structure analysis of tradeoffs; and test robustness of policies and strategies under different socioeconomic and climate futures.

This wide range of decision-support systems and tools is being developed by climate and impact scientists working through university-based research groups, government agencies, private sector consultancies, and NGOs. The tools and systems are available through a number of portals such as the Global Change Information System (see Waple et al., Innovations in Information Management and Access for Assessments, this issue and https://data.globalchange.gov). Federal agencies have also collaborated to develop a US Climate Resilience Toolkit that points users to a range of decision-support tools and systems for assessing vulnerabilities, investigating adaptation options, and appraising risks and costs (https://toolkit.climate.gov). A variety of NGOs also promote information sharing about decision support.

3 The need for evaluation and assessment of decision support

Some decision-support tools and systems are based on sound scientific information and incorporate established principles such as engagement of stakeholders, transparent provision of data and methods, and characterization of uncertainty for decision-making. In others, application of basic principles for sound decision support appears to be lacking, thus creating the need for evaluation and assessment. Some deficiencies are associated with the tools—for example, those that promise temperature and precipitation "forecasts" at a spatial and temporal resolution that is far beyond what is scientifically defensible. In other cases, critical elements of systems associated with improving decision processes are lacking—as examples, the experience of users may have not been considered or evaluated or there may be a lack of transparency regarding assumptions, data, or models used. Most significantly, many decision-support products just use a simple "loading dock" approach that delivers data or methods without considering the context in which they are used. This is partly an issue of ignoring capabilities or needs of users. But, it can also be based on faulty assumptions about the circumstances of decision-making, for example when a decision must be reached; what choices decision-makers can actually select; or how information fits into the broader set of decision-making influences such as political or economic issues, preferences of decision-makers, or other factors related to comfort with and accessibility of information.

Both evaluation and assessment of climate change decision support is needed so that the effectiveness of available systems is not taken for granted. "Evaluation" means data-driven studies of how well a given decision-support system facilitates the use of knowledge to improve decision process or outcomes. Evaluation studies examine specific dimensions of performance, such as communicating uncertainty (for example, Budescu et al. 2012); conveying information, structuring consistent preferences, and mastering the information (for example, Wong-Parodi et al. 2014); or improving interpretability of results and utility for tasks such as scoping or understanding tradeoffs (for example, Parker et al. 2015). Recent research also evaluates deliberate coproduction of knowledge in decision support through different approaches to manage collaboration between scientists and stakeholders (Meadow et al. 2015). The National Aeronautics and Space Administration uses a nine-step Application Readiness Level indicator to evaluate potential decision-support tools in terms of both content and use in the context of end users' decision-making (http://www.nasa.gov/sites/default/files/files/ExpandedARLDefinitions4813.pdf). Evaluation studies are based on data collected through interviews, surveys, experiments (evaluating performance of subjects presented with a task using information from the system), and other methods. "Assessment" of decision support refers to organized processes and reports that synthesize evaluation studies and other research about decision-support processes and systems at a more general level. Assessment of decision-support systems in the NCA would be based on review, comparison, and synthesis of evaluation studies and other sources of information. Assessment outcomes could include identifying elements of good practice and preparation of technical standards and guidelines, comparing effectiveness of different approaches for various purposes, facilitating data collection and additional evaluation, identifying climate and other information needs, and supporting development of improved systems.

4 Adding assessment of decision support to NCA3

The choice to include a decision support chapter in the NCA3 grew out of concern that users of decision-support systems were applying tools and systems that had not been evaluated, creating the potential for a variety of maladaptations and setting up the scientific community for a loss of trust, a worry highlighted in an editorial in Nature magazine (Nature 2010). Preparation of the chapter was facilitated by National Research Council report, *Informing an Effective Response to Climate Change* (Liverman et al. 2010), which synthesized the literature on climate-related decisions and frameworks. Based on this report and a chorus of demands from private and public sector decision-makers, it was apparent that the NCA should initiate an ongoing process in which scientists, intermediaries, and users could interact and begin to inventory and appraise available systems. Over time, data collection and analysis would provide a foundation for rigorous evaluation of what works (or does not) as an input to assessment of decision-support systems. Indeed, one of the key messages of the NCA decision-support chapter is: "Ongoing assessment processes should incorporate evaluation of decision-support tools, their accessibility to decision makers, and their application in decision processes in different sectors and regions" (Moss et al. 2014).

The primary role of this first NCA decision support chapter was not to develop, evaluate, or disseminate specific decision support tools and systems per se. Rather, the role of the NCA was to assess the state of knowledge and practice of decision support in different sectors and regions, thus providing information about good practices that would assist both users and developers of decision-support tools and systems. More specifically, the role of the NCA was to improve communication and promote sustained dialog between users and producers of decision support; facilitate data collection and evaluation of specific systems used in different sectors and regions by researchers; engage professional associations and others in developing good practice guidelines in the context of established professional codes and standards; and, on the basis of this information and the research literature on decision science, assess understanding of current practice and utility in the context of the sectors and regions in which decision-support systems are applied.

The NCA3 decision support chapter is structured around the common conception of adaptive management and decision-making depicted in Fig. 1. The chapter identifies frameworks, tools, methods, and other resources that can assist users as they work through the steps of a typical (idealized) adaptive decision-making process. It describes the types of tools being developed and to includes many examples illustrating current approaches. It does not include evaluation of individual decision-support tools, something that needs to occur in other settings and provide an input to the NCA process. It suggests potential next steps for expanding assessment of decision-support systems in the context of the NCA.

5 Next steps and potential long-term benefits of assessing decision support in the NCA

Including an assessment of decision-support systems in NCA3 was an important first step in positioning the NCA of the future to meet society's evolving and growing needs for climate and impact information for risk management. By example, the decision-support chapter defined a possible role for the NCA in assessing decision-support resources that was consistent with its mandate and available resources. That role provided a framework for beginning to catalog and assess different types of tools and systems offered to help users as they work through the steps of a typical (idealized) adaptive decision-making process. It described the types of tools being developed and included many examples illustrating current approaches. It did not include evaluation of individual decision support tools, something that needs to occur in evaluation research, with the results of these studies providing an input to the NCA process.

The chapter also raised the issue of the next steps the NCA should take as part of the Sustained Assessment process and in future reports. One is to add assessment of decision-support tools and systems to the sectoral and regional chapters of future NCA reports, just as each of these chapters currently includes information on adaptation and mitigation measures relevant to its focus. This will facilitate assessment of the quality of decision support available for different types of decisions and provide a resource for those facing similar issues. Another important step is to continue to include a dedicated chapter on decision support in future NCAs, using the process of preparing the chapter to establish and maintain dialog among scientists, intermediaries, standards organizations, professional societies, consultant organizations, and user groups. The process should include collection of standardized data and information about available systems and how they are applied; synthesis of methods for evaluating decision-support tools and systems; and identification of the elements of a collective effort to promote good practices in decision-support system development, assessment, and use.

It is premature to reach conclusions about the effectiveness of the approach started in NCA3. The potential next steps suggested in the chapter have not yet been reviewed or implemented. But, it may be useful for further consideration of this issue by future NCA leaders and the broader research community to discuss several potential benefits for users, the research community, and decision-support specialists. These include (1) improved understanding of the effectiveness of current decision-support methods; (2) coordinated collection of data and methods for evaluation and research; (3) more clarity on the type of information users need and sources that are useful for informing different types of decisions; and (4) enhanced recognition of the importance of evaluating and assessing both the information content of decision support as well as interfaces, communications, and other aspects of decision-support systems that address issues related to perceptions, cognition, preference formation, deliberation, and other individual and group aspects of decision-making.

Understanding effectiveness and good practice: The NCA3 decision-support chapter included examples of instances in which decision-makers were able to take advantage of scientific information that was thoughtfully developed and communicated in order to improve decision outcomes. Evaluation and assessment are needed to understand what contributes to effectiveness in different settings and at different stages of decision-making. Tools and systems exist to help incorporate scientific information into framing choices, tailoring climate and impact information, accessing relevant information, valuing and comparing outcomes, communicating risk, and other decision-related tasks. The NCA can contribute to a better understanding of good practice, providing guidance to the research community and intermediaries about how to improve their decision support, and educating users on how to differentiate robust from weak information. Assessment can help define standards for decision support. The process will produce more useful results if it engages professional societies and standards organizations that are already involved in promulgating good practices, methods, and data within their areas of specialization. Groups such as the American Society of Heating, Refrigerating, and Air-Conditioning Engineers (ASHRAE), the American Society of Civil Engineers (ASCE), and the National Institute for Building Sciences (NIBS) develop and distribute professional standards, workbooks, and other materials, as well as provide continuing education opportunities. They have an important role to play, and some of them already are participating in the network of NCA-affiliated organizations, NCAnet.

Data and methods for evaluation: Research into human factors that affect how people and organizations interact with scientific information in complex decision environments is essential to improve decision support. An important starting point is establishing baseline conditions (the "before application" condition) and data needs for these kinds of evaluation. There are unresolved research issues regarding how to evaluate whether a decision-making process is improved by a decision-support system. Improvements in process measured by access to better or more timely information does not necessarily lead to better decision-making outcomes. Continued dialog among the research community, intermediaries, and those using decision-support systems is essential for improving evaluation of effectiveness.

Clarifying information needs: Sustained interactions among users, the research community, and intermediaries regarding decision-support needs and practices will be facilitated through the inclusion of assessment of decision support in the NCA. This assessment can identify and convey to decision-support system developers a better understanding of the information needs of users and to users what type and sources of information they should be using. Progress will require moving beyond general discussions by bringing together specific user groups motivated by well-defined decision problems and climate or impact experts who can provide insights about the specific climate and environmental processes of importance to these users. For example, reexamining design loads for buildings and structures in light of changes in exposure (such as intense precipitation or wind speeds) should engage engineers, architects, scientists whose research encompasses the climate processes that affect the relevant exposures, and social scientists who have the capacity to consider how information should be provided to effectively communicate uncertainty.

Examining the human dimensions of decision support: There is a tendency to emphasize tools and products as the critical components of decision-support systems (just as traditional assessments have emphasized reports rather than ongoing dialog). This may be because it is natural, when considering scientific information in decision-making, to focus on the type and quality of the information per se. Indeed, this is critical. But looking at the definition of decision support used in the chapter, "organized efforts to *facilitate the use of knowledge* to improve decision outcomes," it is obvious that the definition of "system" must 1) include the perceptual and social dynamics through which people acquire information; 2) judge its meaning, reliability, and significance; and 3) act on it (or not). A system perspective includes not just the information and delivery mechanisms but how these interact with perceptual, cognitive, and (in some cases) group deliberative processes used to reach decisions. Simply focusing on tools can lead to efforts that provide information, but not in a manner suited to users' mental models or that helps them apply the information consistently and coherently. The NCA process can help reinforce the importance of these factors, by engaging decision analysts, psychologists, and other social scientists trained in research on aspects of decision-making, as well as by explicitly incorporating factors such as knowledge acquisition, information processing, cognition, and group interactions in evaluation of decision-support tools.

6 Increasing usability of NCA findings: assessing uncertainty and communicating confidence

Adding a chapter on decision support was not the only innovation in NCA3 that was intended to increase the relevance of science-based information to users. The NCA also built on previous efforts and updated guidelines for assessing uncertainty and communicating confidence. The intent was to improve transparency for users regarding how confident scientists were about the robustness of their findings and why. This is important because assessment is different from research in that it involves providing information "on demand," which is when decision-makers need it to inform policy, resource management, or planning. In assessments, scientists comb through scientific information to provide an applied synopsis that is relevant to decision-makers; if they are to provide timely input, they cannot wait until 95 % confidence (or some other confidence level) is achieved. Thus, assessments must explicitly describe uncertainties and the level of confidence relevant experts associate with the information they are providing so that users do not misapply it. Studies by Morgan and colleagues (Morgan and Henrion 1990) point to the need to conduct assessments in a self-aware way, so that uncertainty is not underestimated and bias can be identified and analyzed along with an estimate of when and how information could improve. This enables decision-makers to judge for themselves whether to act on available information or wait for scientific knowledge to accrue.

To improve the usability of information in NCA3, guidelines for uncertainty and confidence characterization were developed and provided to authors (Moss and Yohe 2011). They built on previous versions developed for the Intergovernmental Panel on Climate Change (IPCC; Moss and Schneider 2000; Manning and Petit 2003) and the Climate Change Science Program (CSSP; Morgan et al. 2009) and were developed simultaneously with a new set for the IPCC Fifth Assessment Report (Mastrandrea et al. 2011). As in the approach used by the Netherlands Environment Agency (Petersen et al. 2013), a checklist was provided to help the authors complete the process. The steps were to be applied to only the four or so most important conclusions of a chapter and included the following:

1. Frame the most important conclusions with specific questions or uses in mind (rather than just providing "state of science" updates).
2. Evaluate the type, amount, quality, and consistency of evidence as strong, moderate, suggestive, or inconclusive.
3. Formulate well-posed, confirmable conclusions, providing the 90 % confidence range, and high-consequence, low-probability impacts outside that range.
4. Identify key uncertainties and research required to improve knowledge.
5. Assess and report authors' confidence using standard terms (very high, high, medium, and low) and graphics, considering (i) the quality of the evidence (from step 2) and (ii) the level of agreement among subject experts.
6. If evidence is sufficient, provide a likelihood for a well-specified event or impact under a particular scenario, using standardized likelihood ranges (from >9 in 10 chances=very likely to <1 in 10 chances=very unlikely).
7. Prepare a traceable account of the main lines of evidence, uncertainties, and areas of agreement and/or debate among experts to increase transparency.

The guidelines were not implemented exactly as envisioned. Use of the recommended confidence and likelihood terms and ranges was not required in the main body of the report. But, each chapter was required to include traceable accounts for its major conclusions in an appendix, and levels of confidence were included within the traceable accounts. There was feedback that the guidelines were too challenging to apply and authors lacked access to an "uncertainty hotline" function to provide support. The uncertainty guidance drafting team recognized the need for support and in its recommendations, proposed including a decision scientist among the lead authors of key chapters. This recommendation was not implemented. The need for training or support is reaffirmed by the outcome that the guidelines were more likely to be followed when at least one member of the author team felt comfortable with and modeled the process for others. Another complication was that lead authors received guidance on seven additional topics, and conceptual relationships among key issues such as uncertainty, probabilities, scenarios, and risk framing were not established, leading to lack of clarity about what was most important or what to do. Finally, there were concerns among NCA3 leadership that using the confidence terminology (very high, high, medium, and low) in the main body of the report and in synthesis documents would muddy communication of findings, and that including "low confidence" information (even if responding to direct stakeholder needs) had the potential to raise questions of reliability and trustworthiness. In the future, this issue should be thoroughly discussed among leaders, authors, and users.

The tension in the NCA3 between the desire to improve characterization of uncertainty and these general communication impediments did spur at least one positive outcome: a shift in emphasis from "uncertainty," which focuses attention on what is not known, to describing "confidence," which focuses on what is understood in the context of a specific decision. Including "low confidence" categories enabled authors to report relevant information without misleading decision-makers into thinking that information was more certain than it was (a danger of including such information and not reporting a confidence level). Communication scientists need to be more fully engaged to ensure that this does not distract from the NCA's core messages.

Do these problems mean the uncertainty characterization guidance process failed? Some NCA authors reported that the "tedious task" of characterizing confidence and writing the traceable accounts transformed their writing teams from groups of individual experts into assessment teams that agreed upon conclusions describing the current state of knowledge about their topic relevant for different types of decisions. The process also fostered multi- and interdisciplinary thinking, as opposed to individuals remaining solely within their home disciplinary paradigms (see Moser and Davidson 2015). It seems safe to say there were some successes and other areas where implementation was uneven.

In developing the next steps for the Fourth Assessment (NCA4) and the Sustained Assessment, NCA3 lead authors should be surveyed about the process. The traceable accounts need particularly careful review since this part of the guidance was implemented and review could provide valuable information. In what ways are the deliberations of the author teams affected by having to prepare traceable accounts? How useful is the traceable account information that resulted? Much greater effort also needs to be placed on empirical testing (by readers and intended users of information in NCA3) of any language or approach proposed for communicating levels of confidence, verbally, quantitatively, and graphically (Pidgeon and Fischhoff 2011). Ideally, conclusions from this research will inform development of guidelines for NCA4. Making progress in assessing and communicating confidence and uncertainty will also require thinking through relationships between scenarios designated for the assessment, impact model uncertainties, and guidance to authors on related topics such as risk framing.

Finally, more attention should be given to quantitative approaches to uncertainty quantification through use of statistical methods and reduction of uncertainty in climate change projections using advanced experimental design (Katz et al. 2013). In exploring these approaches, it will be important to recognize that many of the most consequential drivers of impacts involve joint probabilities of multiple climate stresses (Tebaldi and Sanso 2009), which introduces added complexity. More sophisticated approaches to use of scenarios in ways better suited to uncertainties in projections could also be explored (for example, Lempert 2013), but would involve significant departures from current assessment practice and thus might best be essayed in the Sustained Assessment process before being brought into a future NCA quadrennial report. Resource and time constraints will challenge application of quantitative approaches, however.

7 Positioning the NCA to meet the information needs of the future

In closing, the NCA process is well situated to advance efforts to facilitate the use of knowledge to improve decision outcomes related to climate change—a grand challenge for society and science in light of rapidly increasing and diversifying impacts. However, meeting these information needs will depend on many factors and advances in research on climate change, impacts, response options, and human factors associated with the use of complex and uncertain scientific information in multifaceted, high-stake decision situations. This paper has reviewed the potential importance of assessment of decision support and improvement of uncertainty characterization and communication of confidence. These improvements, combined with active implementation of the Sustained Assessment process, provides an opportunity to encourage more interactive and focused application of science to decision-making and thus to meet the evolving needs of society for relevant climate science.

References

Bazerman MH (2009) Barriers to acting in time on energy and strategies for overcoming them. Working paper 09-063. Harvard Business School, Cambridge, MA, http://www.hbs.edu/faculty/Publication%20Files/09-063.pdf **Accessed 29 April 2015**

Bierbaum R, Lee A, Smith J et al (2014) Ch. 28: adaptation. In: Melillo JM, Richmond TC, Yohe GW (eds) Climate change impacts in the United States: the third National Climate Assessment. Global Change Research Program, U.S, pp 670–706. doi:10.7930/J07H1GGT

Budescu D, Por H-H et al (2012) Effective communication of uncertainty in the IPCC reports. Clim Chang 113(2):181–200

Buizer JL, Fleming P, Hays SL et al (2013) Report on preparing the nation for change: building a sustained National Climate Assessment process. National Climate Assessment and Development Advisory Committee, Washington DC, http://downloads.globalchange.gov/nca/NCADAC/NCADAC_Sustained_Assessment_Special_Report_Sept2013.pdf **Accessed 20 April 2015**

Buizer JM, Dow K, Black ME et al (2015) Building a sustained climate assessment process. Clim Chang. doi:10.1007/s10584-015-1501-4

Eisenack K, Moser SC, Hoffmann E et al (2014) Explaining and overcoming barriers to climate change adaptation. Nat Clim Chang 4:867–872

Ekstrom JA, Moser SC (2014) Identifying and overcoming barriers in urban adaptation efforts to climate change: case findings from the San Francisco Bay Area, California, USA. Urban Clim 9:54–74. doi:10.1016/j.uclim.2014.06.002

Gifford R (2011) The dragons of inaction: psychological barriers that limit climate change mitigation and adaptation. Am Psychol 66:290–302

Jacoby HD, Janetos AC, Birdsey R et al (2014) Ch. 27: Mitigation. In: Melillo JM, Richmond TC, Yohe GW (eds) Climate change impacts in the United States: the third National Climate Assessment. U.S. Global Change Research, Program, pp 648–669. doi:10.7930/J0C8276J

Katz RW, Craigmile PF, Guttorp P et al (2013) Uncertainty analysis in climate change assessments. Nat Clim Chang 3(9):769–771

Lemos MC, Kirchhoff CJ et al (2012) Narrowing the climate information usability gap. Nat Clim Chang 2(11): 789–794

Lempert R (2013) Scenarios that illuminate vulnerabilities and robust responses. Clim Chang 117(4):627–646

Liverman D, Raven P, Barstow D et al (2010) Informing an effective response to climate change. National Research Council, Washington, DC,
13:978-0-309-14594-7

Manning MR, Petit M (2003) A concept paper for the AR4 cross cutting theme: uncertainties and risk. Intergovernmental Panel on Climate Change, Geneva, Switzerland, Geneva

Mastrandrea M, Field CB, Stocker TF et al (2011) The IPCC AR5 guidance note on consistent treatment of uncertainties: a common approach across the working groups. Clim Chang 108(4):675–691

Meadow AM, Ferguson DB, et al. (2015) Moving toward the deliberate coproduction of climate science knowledge. Weather, Climate, and Society 7(2):179–191

Melillo JM, Richmond TC, Gary W, Yohe GW (eds) (2014) Climate change impacts in the United States: the third National Climate Assessment. Global Change Research Program, U.S. doi:10.7930/J0Z31WJ2

Morgan MG, Henrion M (1990) Uncertainty: a guide to dealing with uncertainty in quantitative risk and policy analysis. Cambridge University Press, Cambridge, New York

Morgan G, Dowlatabadi H, Henrion M et al (2009) Best practice approaches for characterizing, communicating, and incorporating scientific uncertainty in decisionmaking, A report by the Climate Change Science Program and the Subcommittee on Global Change Research. Report series SAP 5.2. National Oceanic and Atmospheric Administration, Washington DC, https://data.globalchange.gov/report/ccsp-sap-5_2-2009 **Accessed 2 April 2015**

Moser SC, Ekstrom J (2010) A framework to diagnose barriers to climate change adaptation. Proc Nat Acad Sci 107(51):22026–22031. doi:10.1073/pnas.1007887107

Moser SC, Davidson MA (2015) The third national climate assessment's coastal chapter: the making of an integrated assessment. Clim Chang. doi:10.1007/s10584-015-1512-1

Moss RH, Schneider SH (2000) Cross-cutting issues in the IPCC Third Assessment Report. In: Pachauri R, Taniguchi T (eds) Uncertainties in the IPCC TAR: recommendations to lead authors for more consistent assessment and reporting. Global Industrial and Social Progress Research Institute for IPCC, Tokyo, pp 33–52

Moss, RH, Yohe G (2011) Assessing and communicating confidence levels and uncertainties in the main conclusions of the NCA 2013 report: guidance for authors and contributors. National Climate Assessment Development and Advisory Committee (NCADAC), Washington, DC. Available at http://www.nesdis.noaa.gov/NCADAC/pdf/nov_16/NCADAC_Mtg_Pres_Nov11_MelMossRichYoh_Final_111611_8b.pdf

Moss R, Scarlett PL, Kenney MA et al (2014) Ch. 26: Decision support: connecting science, risk perception, and decisions. In: Melillo JM, Richmond TC, Yohe GW (eds) Climate change impacts in the United States: the Third National Climate Assessment. U.S. Global Change Research, Program, pp 620–647. doi:10.7930/J0H12ZXG

Nature (2010) Validation required. Nature 463(7283):849

Parker AM, Srinivasan SV et al (2015) Evaluating simulation-derived scenarios for effective decision support. Technol Forecast Soc Chang 91:64–77

Petersen AC, Janssen PHM, van der Sluijs JP, et al. (2013) Guidance for uncertainty assessment and communication, 2nd edn. http://www.pbl.nl/sites/default/files/cms/publicaties/PBL_2013_Guidance-for-uncertainty-assessment-and-communication_712.pdf Accessed 20 April 2015

Pidgeon N, Fischhoff B (2011) The role of social and decision sciences in communicating uncertain climate risks. Nat Clim Chang 1(1):35–41

Tebaldi C, Sanso B (2009) Joint projections of temperature and precipitation change from multiple climate models: a hierarchical Bayesian approach. J R Stat Soc A 172(Part 1):83–106

United States Government Accountability Office (2013) Climate change: future federal adaptation efforts could better support local infrastructure decision makers. Washington, DC. Report series GAO-13-242.

Wong-Parodi G, Fischhoff B et al (2014) A method to evaluate the usability of interactive climate change impact decision aids. Clim Chang 126(3–4):485–493

[illegible] (2011) [illegible]
[illegible]
[illegible]
[illegible]

Climatic Change (2016) 135:157–171
DOI 10.1007/s10584-015-1519-7

Innovations in assessment and adaptation: building on the US National Climate Assessment

Mark Howden[1] **· Katharine L. Jacobs**[2]

Received: 27 March 2015 / Accepted: 20 September 2015 / Published online: 14 October 2015

Abstract Well-targeted scientific assessments can support a range of decision-making processes, and contribute meaningfully to a variety of climate response strategies. This paper focuses on opportunities for climate assessments to be used more effectively to enhance adaptive capacity, particularly drawing from experiences with the third US National Climate Assessment (NCA3). We discuss the evolution of thinking about adaptation as a process and the importance of societal values, as well as the role of assessments in this evolution. We provide a rationale for prioritizing future assessment activities, with an expectation of moving beyond the concept of climate adaptation as an explicit and separable activity from "normal" planning and implementation in the future. Starting with the values and resources that need to be protected or developed by communities rather than starting with an analysis of changes in climate drivers can provide opportunities for reframing climate issues in ways that are likely to result in more positive outcomes. A critical part of successful risk management is monitoring and evaluating the systems of interest to decision-makers and the effectiveness of interventions following integration of climate considerations into ongoing strategic planning activities and implementation. Increasingly this will require consideration of path dependency and coincident events. We argue that climate adaptation is a transitional process that bridges the gap between historically time-tested ways of doing business and the kinds of decision processes that may be required in the future, and that scientific assessments will be increasingly central to these transitions in decision processes over time.

This article is part of a special issue on "The National Climate Assessment: Innovations in Science and Engagement" edited by Katharine Jacobs, Susanne Moser, and James Buizer.

✉ Mark Howden
mark.howden@csiro.au

Katharine L. Jacobs
jacobsk@email.arizona.edu

1 CSIRO Agriculture, GPO Box 1700, Canberra, Australia

2 University of Arizona, Tucson, AZ, USA

1 Introduction

Until recently, the importance of climate assessments as an integral component of adaptation and risk management at multiple scales has not been widely recognized. Climate assessments have undoubtedly played a role in motivating adaptive action, though they have not been explicitly incorporated into ongoing, iterative adaptation processes to the degree they could be (Berrang-Ford et al. 2011). As impacts accumulate and investments are made in risk management, it will be important to document and evaluate the changes in a rigorous, scientific way as well as to evaluate whether or not adaptive actions are successful, in order to prioritize further investments. Because adaptation is a transitional process that bridges the gap between the ways that social, economic, and natural systems have functioned successfully in the past and the decision and management strategies that may be required in the future, scientific assessment at multiple scales will be essential in identifying the vulnerability of and potential impacts to these systems. Further, as the reality of living in a rapidly changing environment becomes a foundation for day-to-day decisions, the idea of adapting *to climate change* per se will necessarily be integrated into decision-making. As that happens, real-time evidence of change caused by the interaction of multiple factors (including climate) will become more ingrained in both policy and management activities. We explore these issues, building from the recent US 3rd National Climate Assessment.

There is already evidence of a transition in federal science programs of the US toward deliberate development of "decision-relevant" climate information to support better risk management and investment decisions. This evidence is found in multiple places, including the most recent 10-year strategic plan of the US Global Change Research Program (USGCRP) (NSTC 2012). This plan incorporates "Informing Decisions," "Conducting Sustained Assessments," and "Communicate and Educate" as pillars of its program in addition to its original (and still dominant) focus of "Advancing Science." There are now interagency coordination activities within USGCRP around each of these pillars, and annual USGCRP budget justifications now include far more discussion of the societal relevance of research investments.

Consistent with this trend, USGCRP's third US National Climate Assessment (NCA3, Melillo et al. 2014) and its associated engagement process were deliberately designed to build long-term adaptive capacity (see Jacobs and Buizer 2015). The NCA3 serves here as a starting point for a broader discussion on the relationships between assessments and adaptation that may be useful in supporting these processes going forward, in both national and international efforts. We evaluate the conceptual underpinning and content of the NCA3 and related studies so as to identify some improvements that could be made to climate assessments to enhance their utility.

2 Terminology: assessments, adaptation, and resilience

Scientific assessments are evaluations of the state of knowledge at a particular point in time and generally include a synthesis of the extant peer-reviewed literature. The NCA3 tried to bring multiple kinds of knowledge, including that of on-the-ground decision-makers, into the assessment process (see Jacobs and Buizer 2015). In part, reaching beyond academia and the federal science agencies is required if assessments are to be effective and relevant in informing climate-related decision-processes. In fact, such assessments need to be transdisciplinary; that is, they must go beyond disciplines to consider decision relevance in specific sectoral and regional applications, accommodate the interactions of physical, social and, economic factors, and be connected to the key issues of concern to stakeholders. These transdisciplinary assessments often require integration of a wide array of sources of knowledge—not just the physical and social sciences, but also the tacit knowledge of citizens, businesses, and local government officials. They also allow testing knowledge of how the "real world" works and the consequences of potential alterations in processes and systems. Because adaptation activities involve applying knowledge, technologies, and management practices in the context of uncertainty (both reducible and irreducible), iteratively testing our understanding of how systems interact and constantly refining the state of knowledge can be an effective way to manage this source of risk and take full advantage of potential opportunities (Cash et al. 2003).

Adaptation, therefore, should be an iterative process that involves ongoing assessments of risks and opportunities based on current and anticipated impacts and vulnerabilities, the development and implementation of response actions and strategies, and the integration of learning from this process into scientific understanding (e.g. Meinke et al. 2009).[1] In recent years the term *resilience* has become very widely used because some perceive it has a more positive connotation than *adaptation.* Some researchers have argued that a resilience approach focuses on maintaining the functioning of valued systems rather than focusing on a return to pre-existing conditions (Gunderson and Holling 2002) or that it helps the process of building adaptive capacity (e.g. Redman 2014), whereas others argue that the term is unhelpful (e.g. Reid and Botterill 2013).

Having considered these and other arguments, we conclude that the term "adaptation" does not inherently invoke a return to status quo and is a more useful way to describe deliberate actions to limit risk or manage opportunities. We also prefer it here because it is more precise: it connotes taking specific outcome-focused action in response to our understanding of past, current, or future conditions. There are various categorizations of adaptive responses (e.g. Smit et al. 2000). Of these, *"proactive" adaptation* is especially dependent on assessment, as it involves anticipating and preparing in a systematic and often transdisciplinary way for future scenarios rather than responding after impacts have occurred. Consequently, this formal approach to adaptation which we address here, appears to be more frequently used by larger, institutional actors (e.g. Tompkins et al. 2010).

Assessing *adaptive capacity* (the knowledge, resources, political/institutional and social capacities needed to manage risk) and adaptation technologies for use in specific applications is an important step in enabling adaptation, but this type of analysis is not commonly incorporated in risk management or adaptation decision processes.[2] *Adaptive management processes* (which depend on ongoing monitoring and iterative learning about the outcomes of alternative strategies) require assessment of changing conditions as well as evaluations of whether adaptation and mitigation activities are having their desired effects.[3] In this case also, the deliberate and rigorous evaluative assessment of options, strategies, and observed outcomes appears to be in its infancy for most economic sectors and geographic locations, but constitutes an important area where future assessments could usefully inform ongoing adaptation processes (Moser and Boykoff 2013; Melillo et al. 2014).

[1] The IPCC WG II Glossary defines *adaptation* as: The process of adjustment to actual or expected climate and its effects. In human systems, adaptation seeks to moderate or avoid harm or exploit beneficial opportunities. In some natural systems, human intervention may facilitate adjustment to expected climate and its effects.

[2] IPCC Glossary, *adaptive capacity*: The ability of systems, institutions, humans, and other organisms to adjust to potential damage, to take advantage of opportunities, or to respond to consequences.

[3] IPCC WG II Glossary, *adaptive management*: A process of iteratively planning, implementing, and modifying strategies for managing resources in the face of uncertainty and change. Adaptive management involves adjusting approaches in response to observations of their effect and changes in the system brought on by resulting feedback effects and other variables.

3 Status of federal adaptation efforts in the United States

In the US, most climate adaptation activities to date have been driven by city, state, industry sector, tribal, and NGO initiatives, rather than by the federal government (NRC 2010). This is not altogether surprising because the impacts of climate and global change are felt locally, and both the impacts and the solutions tend to differ between locations and between sectors. On the other hand, many of these efforts at the local level necessarily involve a broad set of partners, including higher levels of government, and for decades there have been calls for comprehensive federal-level engagement in adaptation and mitigation activities, most recently in recommendations from the National Research Council in the *America's Climate Choices* reports.

To date, political and economic considerations have contributed to a near stalemate of climate response initiatives in the US Congress, with very few programs moving forward. However, since 2010 the Executive Branch of the US government has been facilitating adaptation across the nation. In addition to sponsoring and supporting the NCA3, multiple new programs have been established within and across agencies through regulatory initiatives and by Executive Order. For example, all federal agencies now have emission-reduction goals and adaptation/resilience plans. The President's Climate Action Plan (2013) contains a range of domestic and international emissions management programs and a wide range of programs and partnerships to build resilience, along with Executive Orders in 2009, 2013 and 2014 that itemize specific projects and implementation plans. Agencies must demonstrate how they will protect both their missions and their infrastructure from climate-related impacts, and a number of partnerships have emerged with communities, academia, economic sectors, professional societies, and foundations across the nation. There are now dozens of examples of federal co-investment in local adaptation efforts, particularly in the context of recovery from extreme events. For example, there were explicit efforts to include climate adaptation and preparedness objectives in the capital and social investments made by the federal government and in associated institutional arrangements in the recovery programs for Superstorm Sandy and in the language for the Housing and Urban Development (HUD) National Disaster Resilience Competition.[4] However, differences in goals and power relationships with State and other jurisdictions may restrict the implementation of Federally-supported adaptation measures. The politics of adaptation have to be managed here just as in other contexts (e.g. Vogel et al. 2007; Moser and Ekstrom 2010).

4 The role of the third national climate assessment

The NCA3 was explicitly designed to provide information to support adaptation and mitigation decisions at multiple scales. It is the first US national assessment to contain an evaluation of response strategies in addition to discussions of impacts and vulnerabilities within sectors and regions. Individual chapters cover decision support, connecting science and decision-making, mitigation, adaptation, research needs, and sustained assessment. The report also discusses specific sectoral and region-specific adaptation efforts in most chapters, and contains some references to the connections between assessments and adaptation. Brief references to

[4] https://www.hudexchange.info/cdbg-dr/resilient-recovery/

the importance of ongoing assessments for managing climate risk and maximizing economic opportunities are scattered throughout the NCA3 report. However, the fundamental role of assessment in decision processes is not mentioned explicitly in either the mitigation or adaptation chapters of the report; it is only referred to in the chapter on sustained assessment (Buizer et al. 2013).

The lack of discussion of the relationship between assessments and decision support in the NCA3 synthesis report was primarily due to length limits. However, future assessment processes could be more explicit about the relationship between specific assessment topics and the ways that they either can or should support societal decision processes.

5 Reframing and refining the role of assessments in climate adaptation

5.1 Reframing the conceptual model

The NCA3 Adaptation chapter emphasizes the cyclic or iterative nature of adaptation processes, identifying the importance of stakeholder engagement in the adaptation process. However, portraying a simple learning loop, it implies that assessment inputs occur at only one stage of the adaptation cycle—primarily in the context of evaluating adaptation options. We would argue that assessment activities can and should also be undertaken within the "monitor and evaluate," "identify risks," and "revise strategy" stages. Likewise, the Adaptation chapter focuses primarily on documenting the state of planning and implementing specific adaptation actions at multiple scales without substantial emphasis on the state of knowledge of adaptation processes more generally, likely related to the relative infancy of evaluating adaptation efforts.

There are more detailed depictions of adaptation approaches. In particular, the adaptation cycle of Willows and Connell (2003) importantly identifies an iterative risk assessment sub-cycle. Another approach, outlined by Meinke et al. (2009), identifies multiple points of assessment and incorporates initial components of a theory of change. We suggest yet another alternative schematic (Fig. 1) that draws on these earlier figures but more explicitly illustrates how multiple assessment types/stages support adaptation.

In this figure, the central triangle draws a link to Bennett's (1975) hierarchy of objectives, recognizing that a range of societal 'values'—though clearly evolving over time and visible in multiple ways through policy, regulations, economic choices, and lifestyles—form the basis for societal decision-processes that affect everything else in the cycle (O'Brien 2009). Emerging from these values are aspirations that must be tempered by what can actually be done due to physical, economic, social and psychological constraints (e.g. Grothmann and Patt 2005). Above these in the triangle are strategies to realize the aspirations. Specific adaptation decisions are then needed to take the strategies into adaptation actions.

Surrounding all and interacting particularly with the adjacent parts of the triangle is the adaptation-assessment cycle, initiated (at right) with scoping (i.e. should we even be concerned about climate change risks?), and proceeding through an impacts assessment/adaptation assessment cycle of traditional "risk assessment" or "vulnerability assessment," then an assessment of response strategies, necessary associated capacities, and barriers, followed by assessment of effectiveness of the actions taken. This evaluation work should ideally be done

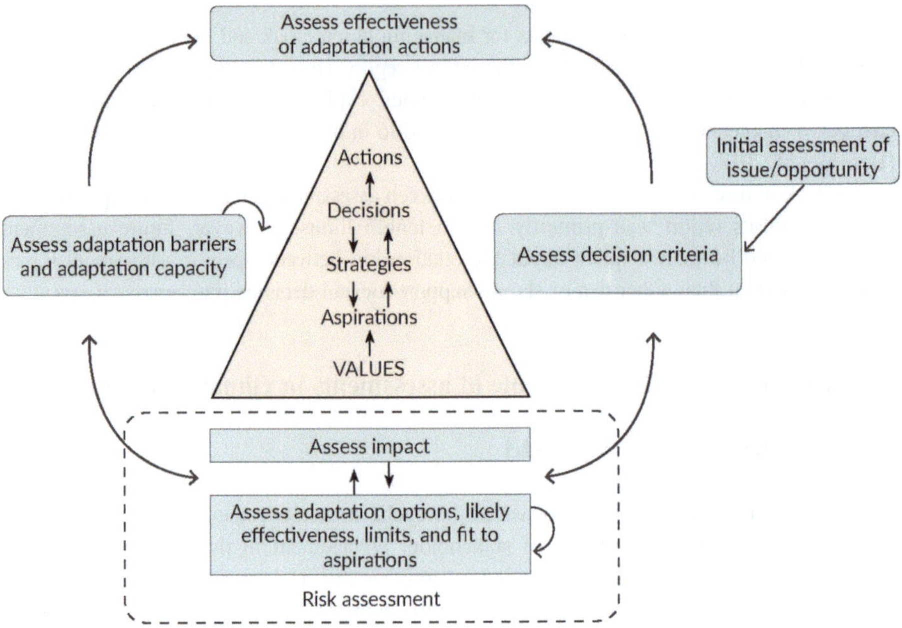

Fig. 1 Conceptual outline of the roles of assessment in adaptation

after metrics are selected, baseline monitoring processes are in place, and the actions are implemented.

5.2 Assessment of barriers and adaptive capacity

Evaluation of institutional barriers and adaptive capacity is a particularly important consideration in successful adaptation and there is a growing literature on building both specific and generic climate adaptive capacity as a pre-disposing factor for adaptation (Lemos et al. 2013; IPCC 2014a). There are also several ways in which adaptive capacity can be assessed, at least in relative terms, at individual, community, regional, and national levels (Nelson et al. 2010) that can inform decision makers about where adaptive capacity is weak or strong and what can be done to build it. It is important, however, to recognize the limitations of large-scale (national) or generalized adaptive capacity assessment approaches in terms of how they are conceptualized, and the different approaches needed, depending on the specific climate risks being considered and their utility in actual decision-making (Hinkel 2011). Adaptive capacity can be more usefully assessed in systems that are more narrowly defined, where deductive approaches can identify key adaptive capacity variables. Inductive approaches can link these to specific outcomes where data are consistently available (Hinkel 2011).

While in the NCA3 there are many statements recognizing the need for enhancing adaptive capacity, there are only two instances where there is any detail about what constitutes adaptive capacity and what aspects needed to be enhanced. Notably, improving understanding of adaptive capacity was recognized as a key research need in the NCA3 Research Agenda chapter. Subsequent assessment reports could focus more explicitly on assessing specific and general adaptive capacity within specific decision contexts in order to facilitate adaptation (Lemos et al. 2013).

5.3 Intersections of adaptation and mitigation

Despite its advantages, Fig. 1 does not consider the impacts of adaptation actions on emissions trajectories (and vice-versa). These considerations should be integral but are often missing in adaptation assessments, and this omission may become more problematic as climate impacts escalate. There is a particular need to assess how adaptation may affect achievement of mitigation goals and how mitigation activities may limit or facilitate adaptation over time (Rosenzweig and Tubiello 2007) rather than treating these as independent topics. For example, the 2014 IPCC Working Group II Report (IPCC 2014a) states that most categories of adaptation to climate change have positive impacts on mitigation. These include healthier and more productive soils, crops, and grasslands, greater water security and protection from flooding, greater diversification of crops and agroforestry to cope with more extreme weather events and climate change. For example, from 2002 to 2007, Uganda increased the number of certified organic farmers by 359 %, increasing prices received and organic exports, while greenhouse gas emissions for these farms are estimated to be 64 % lower than those from non-organic farms (Sukhdev et al. 2010).

Because information for decision support is becoming more critically needed over time, and there are so many potential paths forward, assessments should have clear strategic framing focused on achieving specific objectives (NRC 2007). Developing assessments that are truly useful is likely to require moving away from the linear, climate-impact focused assessment process into assessments that are more decision-focused and iterative and that include both adaptation and mitigation components, and with multiple scales of analysis "nested" within a broader framework as suggested by the NRC 2007 report. These more focused, but perhaps more intellectually-challenging kinds of assessments could be periodically harvested as "interim reports" in a sustained assessment process as was suggested in the NCA3 advisory committee report on building a sustained assessment process (Buizer et al. 2013). Development of reports that are responsive to the scale, context, and decisions at hand, but that focus on synergies and trade-offs (including with mitigation goals) could be another important output of a sustained assessment.

5.4 Reframing climate issues: implications for assessments and adaptation

Accumulating changes in global temperatures and other climate-related factors are increasingly being attributed to human influence via net greenhouse gas emissions rather than the natural processes that have controlled past climate fluctuations, such as solar and volcanic activity, changes in the Earth's orbit and decadal variability driven by ocean conditions (Melillo et al. 2014; IPCC 2014b; Kokic et al. 2014). In contrast, recent public surveys in the US show a majority of the populace acknowledges that climate means and extremes are changing but many do not accept that these changes are due to human activity, instead attributing them to climate variability (Moser 2014), presumably one of the more persistent, long-term causes of variation noted above. Taken at face value, a lack of attribution to human causes implies an expectation that current climate trends will reverse at some stage, returning to 'normal'. This has direct and negative implications for adaptation implementation, especially for those adaptations with long lifetimes. Because of these fundamental disconnects between scientific observations and public opinion, evaluation of alternative values-based paths to risk management (Fig. 1) is a useful strategy, especially in cases where there is reluctance to explicitly acknowledge that the climate is changing.

Humans are more or less adapted to the climate of the recent past but existing risk management efforts are often inadequate, creating a widely recognized "adaptation deficit" (Burton 2009) that is not restricted to less developed nations. For example, with billions of dollars of damage incurred annually from floods, drought, storm surges, and heat waves, it is hard to argue that the US is fully prepared even for current climate conditions. The potential for increased damage is widely recognized, particularly along coasts, yet there continue to be investments in new infrastructure, private property, and businesses that will not withstand the highest current tides, much less future, climate-related, extremes. Similarly, there are often vulnerable groups within developed nations (e.g. many Native American communities), which are less prepared for the impacts of climate change (see Maldonado et al. 2015).

Despite obvious limitations, one option to enhance risk management efforts without explicitly incorporating additional risks of anthropogenic change is using historical extremes as a proxy for what the future might bring (e.g. Hallegatte 2009). This can offer significant benefits in adaptation efforts if, due to perception issues, "climate change" is not an acceptable entry point. Additionally, this can be inclusive of welfare of affected groups and flow-on impacts on systems of concern which may facilitate policy implementation (Dupuis and Knoepfel 2013). Historical and paleo-records (including long-term tree ring, ice core, and other proxy data) provide an estimate of the envelope of variation in key climate-related extremes that enables people to relate their own experience to a broader range of possible events without challenging their underlying perspective on anthropogenic climate change. Similarly, using space for time analogues (i.e. looking for locations that have historically had the basic climate characteristics anticipated in the future) can be another way of effective engagement, as the risk management approaches for that climate have already been addressed, facilitating adoption without the trial-and-error that would otherwise have to occur (Dunn et al. 2015).

Limitations to the above approaches include expectations of change in both the return period and the intensity of some events. For example, a range of studies show return periods for time-specified events shrinking dramatically (IPCC 2014b). Experience in other nations, such as Australia, shows that preparedness for events beyond the envelope of historical experience is critical to protecting communities and ecosystems from devastating impacts (IPCC 2014a). And in the case of flooding caused by sea-level rise in the conterminous US, Tebaldi et al. (2012) found that by 2050 what was historically a 1-in-100 year event may become an annual event in some locations. Similarly, the magnitude of some extremes (e.g., hurricanes, downpours, heat waves) is changing and is expected to intensify (Melillo et al. 2014; IPCC 2014b). Clearly, if these analyses are sound, then depending only on historical data and impact relationships can lead to inadequate risk assessment (especially relating to coincident events) and poor formulation of adaptation options. Figure 2 illustrates the implications of assuming no significant change in the probability distributions of climate risk: the portion of land experiencing >3 sigma summer heat in a given year increased from 0.1 to 0.2 % in 1951–1980 to 10 % in 2001–2011 (Hansen et al. 2012).

Further progress in climate change attribution is needed to improve our ability to separate the influences of climate variability from the climate change trend, because this will provide better predictive capacity and reduce uncertainties. Nevertheless, in some situations it is possible to reduce risk without necessarily needing more accurate predictions of future conditions through the development of no-regret, low-regret or robust adaptations (e.g. Lempert 2013). Again the importance of assessments is emphasized: without ongoing

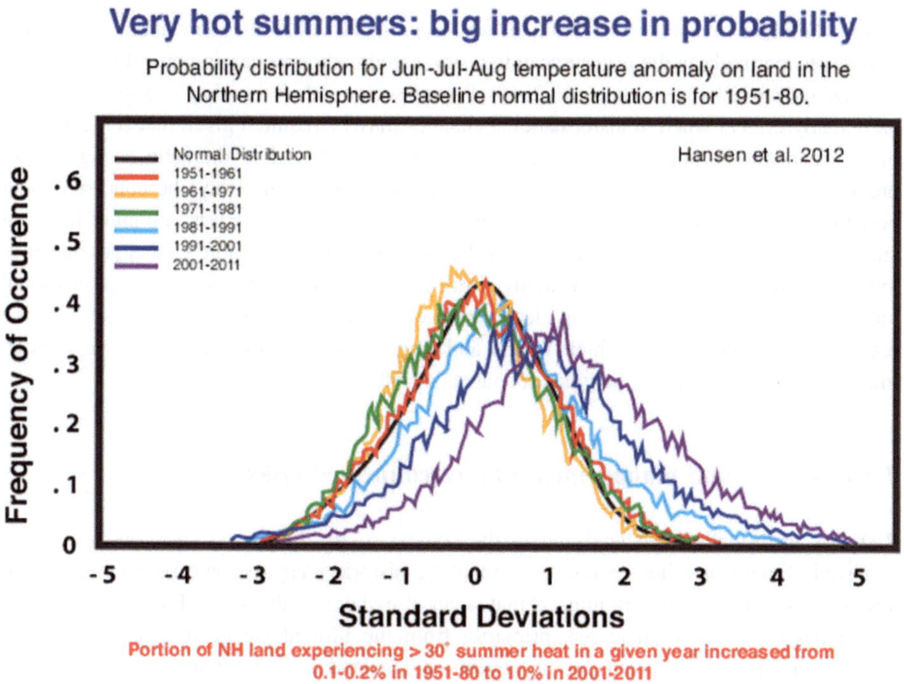

Fig. 2 Probability distribution for June-July temperature anomaly on land in the Northern Hemisphere. Baseline normal distribution is for 1951–1980. Hansen et al. (2012)

evaluations of what we know and how we know it, and how that relates to specific decisions, there will be little progress in managing changing risks. Similarly, personal experience of climate extremes is a key factor affecting public opinion about climate change (Brulle et al. 2012) and being able to provide robust attribution of the human influence on these extremes will likely be a key change factor in future public acceptance of climate change.

In the meantime, by ignoring the directional change component in climate, there is a risk of a growing gap between 'current best practice' and 'past practice' as it relates to management of climate for communities, land use, industry and ecosystems—resulting in either underperformance or unnecessary risk.

6 Assessment, adaptation and the increasing role of non-government actors

If deviation of the climate from past patterns increases as anticipated, there may be an increasing need for more transformational adaptations that involve significant changes to human systems (Park et al. 2012). Of note, such dramatic changes may provide opportunities for changing roles of non-governmental organizations (NGOs), academia and the private sector in decisions and investments associated with adaptation. These different actors often request or seek new and different kinds of information to support these decisions (Dowd et al. 2014).

Governments have often been the insurers of last resort, stepping in after large-scale disasters to subsidize recovery processes and to support more transformational change. But government actors are becoming more hamstrung at multiple scales in many regions and sectors, limited by resource availability as well as political will. Some foundations, NGOs and businesses, particularly the re-insurance industry, are already encouraging proactive decisions that will reduce risk (Tompkins et al. 2010). Private sector and civil society engagement seems particularly critical where transformational changes may be required given that some government organizations appear to be averse to decisions that lead to dramatic changes from previous 'best practices' (Park et al. 2012). However, there will always be concerns about how the private sector's interests can be melded with the public's interests in light of financial and regulatory incentive systems in which they operate. There are more challenges ahead in managing and assessing risk in a highly distributed, multi-actor context where professional judgement and multiple kinds of evidence are used. A key challenge in an effective adaptive response is understanding which trends and impacts to monitor, because there is such imperfect understanding of the greatest sources of risk.

7 Assessment and management of interconnected risks

Risk has historically been analyzed in the context of the likelihood and consequence of localized climate or other events. Unfortunately, historic analyses have sometimes missed key sources of risk, including impacts that occur elsewhere on the globe. For example, many companies evaluated climate risk primarily from the limited perspective of risk to their physical facilities, until recent flooding events in Thailand made the issues of global supply chain risk very clear (Wai and Wongsurawat 2009; Liverman 2015).

In addition to the interconnectivity of risks across distances, there are cascading effects of risks across multiple sectors, due to the many lifelines that support proper functioning of communities. The aftermath of hurricanes Katrina and Rita and of Superstorm Sandy illustrated the interconnected impacts well beyond the immediate impacts of the climatic event, such as overloaded health care systems, breakdowns of transportation and communications systems, and disruptions of critical water supply, water treatment, and electric power systems (Melillo et al. 2014). These cascading effects can often be anticipated but frequently are not.

It is impossible to anticipate all sources of risk and even when they are anticipated, there can be a range of rational reasons why effective risk management is not implemented. Hence, there is a need for solutions that help limit the worst of possible impacts, and much more sophisticated ways to formally assess their utility under a variety of future conditions. Help in identifying which risks to focus on can be derived from targeted and more sophisticated assessment processes that support risk management efforts and through strategic planning processes for individual businesses, ecosystems, sectors, or communities that better include climate risk and consideration of ideas such as robust adaptation and systems that are safe-to-fail (e.g. Ahern 2011). But perhaps even more important is awareness of the underlying values and institutions that lead to our existing decision processes and priorities.

These issues are made much more cumbersome by unclear and sometimes overly academic definitions and disciplinary approaches to analysis (Hinkel 2011). The NCA3 authors as well as managers in the "real world" struggle with issues of language, perception, statistics and communication related to risk. But it is clear that a well-structured assessment process—especially an ongoing and strategically focused effort to explore how risks change over time—can help define risks in a meaningful way and help evaluate alternative paths forward for managing them (e.g. Lonsdale et al. 2008). Working directly with affected decision-makers and stakeholders on these issues adds both complexity and richness to these discussions—and is a cornerstone of developing useful and implementable strategies (Cash et al. 2003).

8 Paths forward for assessments and adaptation

One approach to dealing with complex intersections is, as illustrated in Fig. 1, starting the discussion with identifying the values that stakeholders want to protect or develop rather than with the expected changes in the climate system (e.g. van Aalst et al. 2008). Compared with more top-down assessments, this alternative framing incorporates social science strategies and techniques from the outset rather than being initiated after assessment of the biophysical impacts. Pursuing adaptation by starting with values and aspirations may require triple-loop learning [(Swieringa and Wierdsma 1992: starting with a first loop focused on what to do (rules); followed by a second loop: learning what to do (developing insights); and finally pursuing the third loop: learning how to learn (principles)]. One well-established approach that can support this type of learning is testing the reaction of decision-makers to alternative "virtual" outcomes through participatory scenario development linked into adaptation planning efforts to help with the transition from largely reactive responses to events to more anticipatory paths and approaches. However, there can be significant policy and political dimensions, often related to incumbency, that can act to limit or influence these processes (e.g. Vogel et al. 2007; Moser and Ekstrom 2010).

Understanding the acceptability of different levels and types of risk to resources of value to specific decision-makers and discussing the implications of potential disruptions to a range of communities and systems helps put the costs and benefits of adaptation actions into perspective. In particular there is a potential need for the equivalent of business continuity planning for communities, ecosystems, and institutions. Our perspective is that the current, deliberate overlay of climate risks over existing planning processes is a "transitional" approach to planning that bridges the gap between time-tested ways of doing business and the kinds of decision processes that may be required in the future. For risk management to be truly effective, the concept of change must be integrated into our psyches in a much more effective way and incorporated into the planning and implementation of relevant activities: those activities that are designed to maintain the *status quo* as well as those that are intended to transform systems to a more robust future state.

We recommend a transition to embedding or mainstreaming information about future climate conditions and associated stresses in decisions just as other kinds of risks and considerations are managed: by evaluating implications for processes, outcomes, or objectives of decisions and planning, rather than depending on independent climate change assessments. This approach has frequently been raised in the literature but it may take time for this to become both an explicit and implicit component of most planning and implementation and there are likely to be many constraints to be negotiated (e.g. Moser and Ekstrom 2010 and see Section 9). As climate considerations have so many subtle repercussions, integrating changing climate conditions into strategic planning considerations is a challenge, but it must become ingrained in our decision-making. There is however, a counterargument in the short term: mainstreaming changing climate conditions into planning processes without accounting for political economy and vested interests could result in diminishing attention being paid to the topic.

Although the US is only just beginning to explicitly incorporate climate considerations in general decision processes, there are many examples of progress. For example, new Federal

Emergency Management Agency flood maps and flood insurance policies require incorporation of "freeboard" above current 100-year flood levels (based on expected sea-level rise by mid-century) into coastal property insurance. The American Society of Civil Engineering is working to incorporate climate change considerations into its engineering standards. Ongoing assessments of the effectiveness of these and other adaptation efforts will allow evaluation of progress in reducing risk and in embracing new planning and adaptation paradigms.

9 Future assessment/information needs for adaptation

As future assessments are designed and data are collected to facilitate evaluation of the effectiveness of adaptation efforts, it will be important to consider when and how adaptation decisions can most effectively be made at multiple scales and the information needs for evaluating the "ripeness" of information for supporting short versus long-term decisions. This will require a strategic approach to assessment, with a range of products developed over time (Buizer et al. 2013). Timing factors also need to be considered in planning for, paying for, implementing, and assessing adaptation effectiveness. There are many technical challenges, including relatively few sources of information that can resolve the data needs of specific categories of stakeholders, difficulties in scaling projects up or down over time, and lags between decisions, investments, and benefits, especially when preparing for high-impact, low-probability events. Likewise there can be social and cultural challenges, particularly where adaptation options have significant trade-offs or are conflictual in nature because of asymmetric costs and benefits.

Path dependency is another important topic that needs attention when prioritizing adaptation efforts. Analysis is needed that supports robust decision-making, the timing of decisions and implementation relative to rates and sources of increasing risk, and optimizing across a range of possible decisions at multiple scales. Future assessments could usefully address both path dependencies and adaptation pathways, i.e., a more dynamic conception of adaptation in the face of a continually changing climate.

10 Conclusions

This paper explores the multiple ways that assessments can support adaptation activities, and provides some new perspectives on adaptation processes themselves. It provides a rationale for prioritizing future assessments, with an expectation of moving beyond the concept of climate adaptation as an explicit and separable activity from "normal" planning and implementation activities in the future. We suggest paths forward for practitioners who are interested in managing risk, including in circumstances where climate change is not widely accepted as a concept. Some of the options include supporting exploration of alternative futures through a focus on extreme events of the past, and enhancing documentation of pre-adaptation (historical) baselines from which to evaluate progress and allow learning.

Moving beyond climate drivers as the initiation point for adaptation and working instead to identify the values and resources that need to be protected or developed in communities may require a new set of assessment tools for managing risk, but it may also be a more useful conceptual framing as the pace of multiple components of change continues to accelerate. Engaging in sustained monitoring of what is changing and whether adaptation actions are effective in light of established values and goals is the logical next step, along with integrating climate considerations into ongoing strategic planning activities and a wide range of decisions about human systems. A more sophisticated use of decision analysis may also be required, including assessing and understanding both actual and perceived risks in specific decision contexts. In each case there is an explicit role for formal assessments in enhancing adaptation processes.

Acknowledgments We thank Mary Black (University of Arizona) for her assistance in preparing the manuscript.

References

Ahern J (2011) From fail-safe to safe-to-fail: sustainability and resilience in the new urban world. Landsc Urban Plan 100:341–343

Bennett C (1975) Up the hierarchy. J Extension [On-line], 13(2). Available at: http://www.joe.org/joe/1975march/75-2-a1.pdf

Berrang-Ford L, Ford JD, Paterson JA (2011) Are we adapting to climate change? Glob Environ Chang 21(1): 25–33

Brulle RJ, Carmichael J, Jenkins JC (2012) Shifting public opinion on climate change: an empirical assessment of factors influencing concern over climate change in the U.S., 2002–2010. Clim Chang 114(2):169–188

Buizer, JL, Fleming, P, Hays SL, Dow K, Field CB, Gustafson D, Luers A, Moss RH (2013) Preparing the nation for change: Building a sustained National Climate Assessment process, National Climate Assessment and Development Advisory Committee

Burton I (2009) Climate change and the adaptation deficit. In: Schipper ELF, Burton I (eds) The Earthscan reader on adaptation to climate change. Earthscan, London

Cash DW, Clark WC, Alcock F, Dickson NM, Eckley N, Guston DH, Jager J, Mitchell RB (2003) Knowledge systems for sustainable development. Proc Natl Acad Sci U S A 100:8085–8091

Dowd AM, Marshall N, Fleming A, Jakku E, Gaillard E, Howden M (2014) The role of networks in transforming Australian agriculture. Nat Clim Chang 4:558–563

Dunn M, Lindesay J, Howden M (2015) Spatial and temporal scales of future climate information for climate change adaptation in viticulture: a case study of user needs in the Australian wine-grape sector. Aust J Grape Wine Res 221:165–179

Dupuis J, Knoepfel P (2013) The adaptation policy paradox: the implementation deficit of policies framed as climate change adaptation. Ecol Soc 18(4):31

Executive Office of the President (2013) The President's Climate Action Plan, Washington, DC, June 2013. http://www.whitehouse.gov/sites/default/files/image/president27sclimateactionplan.pdf

Executive Order 13514, Federal leadership in environmental, energy, and economic performance, Oct. 5, 2009, Federal Register 74, 52117

Executive Order 13653, Preparing the United States for the impacts of climate change, Nov. 6, 2013, Federal Register 78 (215), 66819–66824

Executive Order 13677, Climate-Resilient International Development, Sept. 26, 2014, Federal Register 79 (58231)

Grothmann T, Patt A (2005) Adaptive capacity and human cognition: the process of individual adaptation to climate change. Glob Environ Chang 15:199–213

Gunderson LH, Holling CS (2002) Panarchy: understanding transformations in human and natural systems. Island Press

Hallegatte S (2009) Strategies to adapt to an uncertain climate change. Glob Environ Chang 19:240–247

Hansen J, Sato M, Ruedy R (2012) Perception of climate change. Proc Natl Acad Sci 109:E2415–E2423. doi:10.1073/pnas.1205276109

Hinkel J (2011) Indicators of vulnerability and adaptive capacity: towards a clarification of the science policy interface. Global Environ Chang 21:198–208

IPCC (2014a) Climate change 2014: impacts, adaptation, and vulnerability. Part A: Global and sectoral aspects. Contribution of Working Group II to the Fifth Assessment Report of the Intergovernmental Panel on Climate Change [Field CB, Barros VR, Dokken DJ, Mach KJ, Mastrandrea MD, Bilir TE, Chatterjee M, et al. (eds.)]. Cambridge University Press

IPCC (2014b) Climate Change 2014: Synthesis report. Contribution of Working Groups I, II and III to the Fifth Assessment Report of the Intergovernmental Panel on Climate Change [Core Writing Team, RK Pachauri and LA Meyer (eds.)]. Geneva, Switzerland
Jacobs KL, James LB (2015) Building community, credibility and knowledge: the third U.S. National Climate Assessment. Clim Chang. doi:10.1007/s10584-015-1445-8
Kokic P, Crimp S, Howden M (2014) A probabilistic analysis of human influence on recent record global mean temperature changes. Clim Risk Manag 3:1–12
Lemos MC, Agrawal A, Johns O, Nelson D, Engle N (2013) Building adaptive capacity to climate change in less developed countries. In: Asrar GR, Hurrell JW (eds) Climate science for serving society: research, modeling and prediction priorities. OSC Springer Science, Dordrecht, pp 437–458
Lempert R (2013) Scenarios that illuminate vulnerabilities and robust responses. Clim Chang 117(4):627–646
Liverman D (2015) U.S. National climate assessment gaps and research needs: overview, the economy and the international context. Clim Chang. doi:10.1007/s20584-015-1464-5
Lonsdale KG, Downing TE, Nicholls RJ, Parker D, Vafeidis AT, Dawson R, Hall J (2008) Plausible responses to the threat of rapid sea-level rise in the Thames Estuary. Clim Chang 91:145–169
Maldonado J, Bennett TMB, Chief K, Cochran P, Cozzetto K, Gough B, Redsteer MH, Lynn K, Maynard N, Voggesser G (2015) Engagement with indigenous peoples and honoring traditional knowledge systems. Clim Chang. doi:10.1007/s10584-015-1535-7
Meinke H, Howden SM, Struik PC, Nelson R, Rodriguez D, Chapman SC (2009) Adaptation science for agriculture and natural resource management - urgency and theoretical basis. Curr Opin Environ Sustain 1:69–76
Melillo JM, Richmond TC, Yohe GW, eds. (2014) Climate change consequences in the United States: The Third National Climate Assessment, U.S. Global Change Research Program (https://nca2014.globalchange.gov)
Moser SC (2014) Communicating climate change adaptation: the art and science of public engagement when climate change comes home. Wiley Interdiscip Rev Clim Chang 5:337–358
Moser SC, Boykoff M (eds) (2013) Successful adaptation to climate change: linking science and practice in a rapidly changing world. Routledge, London
Moser SC, Ekstrom JA (2010) A framework to diagnose barriers to climate change adaptation. Proc Natl Acad Sci U S A 107:22026–22031
National Research Council (2007) Board on atmospheric sciences and climate. Analysis of global change assessments: lessons learned. National Academies Press, Washington, DC
National Research Council (2010) Panel on adapting to the impacts of climate change america's climate choices. adapting to the impacts of future climate change. National Academies Press, Washington, DC
National Science Technology Council (2012) National global change research plan 2012–2021: A strategic plan for the U. S. Global Change Research Program
Nelson R, Kokic P, Crimp S, Martin P, Meinke H, Howden SM, de Voil P, Nidumolu U (2010) The vulnerability of Australian rural communities to climate variability and change: part II—integrating impacts with adaptive capacity. Environ Sci Pol 13:18–27
NSTC 2012: http://downloads.globalchange.gov/strategic-plan/2012/usgcrp-strategic-plan-2012.pdf
O'Brien, KL (2009) Do values subjectively define the limits to climate change adaptation? In: Adger WN, Lorenzoni I, O'Brien KL (eds) Adapting to climate change: thresholds, values, governance. Cambridge University Press, pp 164–180
Park SE, Marshall NA, Jakku E, Dowd AM, Howden SM, Mendham E, Fleming A (2012) Informing adaptation responses to climate change through theories of transformation. Glob Environ Chang 22:115–126
Redman CL (2014) Should sustainability and resilience be combined or remain distinct pursuits? Ecol Soc 19:37. doi:10.5751/ES-06390-190237
Reid R, Botterill LC (2013) The multiple meanings of resilience: an overview of the literature. Aust J Publ Admin 72:31–40
Rosenzweig C, Tubiello FN (2007) Adaptation and mitigation strategies in agriculture: an analysis of potential synergies. Mitig Adapt Strat Glob Chang 12:855–873. doi:10.1007/s11027-007-9103-8
Smit B, Burton I, Klein RJT, Wandel J (2000) An anatomy of adaptation to climate change and variability. Clim Chang 45:223–251
Sukhdev, P, Stone S, Nuttall, N (2010). Green economy developing countries success stories. UNEP. pp 15
Swieringa J, Wierdsma A (1992) Becoming a Learning Organization: beyond the learning curve. Addison-Wesley, Wokingham
Tebaldi C, Strauss BH, Zervas CE (2012) Modelling sea level rise impacts on storm surges along US coasts. Environ Res Lett 7(1), doi:10.1088/1748-9326/7/1/014032
Tompkins EL, Adger WN, Boyd E, Nicholson-Cole S, Weatherhead K, Arnell N (2010) Observed adaptation to climate change: UK evidence of transition to a well-adapting society. Glob Environ Chang 20:627–635
Van Aalst MK, Cannon T, Burton I (2008) Community level adaptation to climate change: the potential role of participatory community risk assessment. Glob Environ Chan 18:165–179

Vogel C, Moser S, Kasperson R, Dabelko GD (2007) Linking vulnerability, adaptation, and resilience science to practice: pathways, players, and partnerships. Glob Environ Chang 17:349–364

Wai LC, Wongsurawat W (2009) Crisis management: Western digital's 46-day recovery from the 2011 flood disaster in Thailand. Strateg Leadersh 41(1):34–38

Willows R, Connell R (2003) Climate adaptation: risk, uncertainty and decision-making. UKCIP Technical Report, UK Climate Impacts Programme, Oxford

Climatic Change (2016) 135:173–186
DOI 10.1007/s10584-015-1464-5

U.S. National climate assessment gaps and research needs: overview, the economy and the international context

Diana Liverman [1]

Received: 27 December 2014 / Accepted: 3 July 2015 / Published online: 24 July 2015

Abstract A number of knowledge gaps and research priorities emerged during the third US National Climate Assessment (NCA3). Several are also gaps in the latest IPCC WG2 report. These omissions reflect major gaps in the underlying research base from which these assessments draw. These include the challenge of estimating the costs and benefits of climate change impacts and responses to climate change and the need for research on climate impacts on important sectors such as manufacturing and services. Climate impacts also need to be assessed within an international context in an increasingly connected and globalized world. Climate change is being experienced not only through changes within a locality but also through the impacts of climate change in other regions connected through trade, prices, and commodity chains, migratory species, human mobility and networked communications. Also under-researched are the connections and tradeoffs between responses to climate change at or across different scales, especially between adaptation and mitigation or between climate responses and other environmental and social policies. This paper discusses some of these research priorities, illustrating their significance through analysis of economic and international connections and case studies of responses to climate change. It also critically reflects on the process of developing research needs as part of the assessment process.

1 Introduction

The third United States National Climate Assessment (NCA3) and the 5th and latest report of the Intergovernmental Panel on Climate Change (IPCC) Working Group 2 are important benchmarks in our understanding of the impacts of climate change and how we might respond. They synthesize hundreds of studies of how climate change is affecting the Earth, socioeconomic systems, and major regions. Yet, even after five IPCC reports and three US Global

This article is part of a special issue on "The National Climate Assessment: Innovations in Science and Engagement" edited by Katharine Jacobs, Susanne Moser, and James Buizer.

✉ Diana Liverman
liverman@email.arizona.edu

[1] University of Arizona, Tucson, AZ, USA

Change Research Program (USGCRP) climate assessments there are still major gaps in research such that the overall picture of how climate change is affecting us is incomplete or inaccurate and this may result in underestimates and inefficiencies in our response to climate change.

The research agenda chapter of the National Climate Assessment (Corell et al. 2014) together with the Sustained Assessment chapter (Hall et al. 2014) and the Sustained Assessment Report (Buizer et al 2013) identified a number of these research gaps and priorities based on input from other chapters, technical inputs to the NCA3 and the NCA advisory committee. These gaps and research needs are identified in the first section of this paper together with some guidelines for setting priorities for assessments.

These research gaps are then explored in greater depth including the need to study and assess the full range of climate impacts on the economy and the importance of analyzing local impacts in the international context of a globalized economy. Assessing a wider range of impacts on the economy is an important step in informing decision makers and the public about the climate impacts that may affect the sectors – services and manufacturing – that drive most regional and national economies in terms of GDP and employment. Understanding local impacts in an international context is critical if the research and policy community want to understand how climate change can affect local costs of energy, food, water and other goods. In an increasingly globalized world, with few trade barriers, many people and regions are dependent on imported goods or on goods whose prices are set by global markets. Whether it is the costs of computer components or oil or the price of winter fruit and vegetables, companies and consumers in the US and elsewhere are likely to be affected by climatic changes and extremes in other regions.

2 Research needs and the national climate assessment

Several methods were used to develop the research needs chapter of the NCA. Lead authors for each chapter in the assessment were asked to provide a summary of the research gaps and priorities identified in their chapters. The research needs chapter authors also consulted the final chapter drafts for any material relevant to research needs. They also used recent reports from the NRC and USGCRP as a background for the chapter, and presented the list of research needs and criteria for priorities to two meetings of the 60 member NCADAC (National Climate Assessment and Development Advisory Committee) for comment. As with other NCA chapters the chapter was extensively reviewed and revised in response to comments.

The research agenda chapter of the NCA identifies the importance of activities supporting assessment such as research to observe and understand the climate system, improve understanding of climate impacts, vulnerability and adaptation pathways, identify mitigation options and improve capacity and decision support (Table 1). As such, it directly responds to the legal mandate to not only assess and evaluate what is known, but also to discuss uncertainties (USGCRP 1990, Sec. 106). Of course, many of these priorities have been identified in other reports (e.g., NRC Committee on America's Climate Choices 2010) and most are included in the US Global Change Research Program's Strategic Plan (USGCRP 2012).

Within the goals and capabilities listed in Table 1 there are several research needs that emerged strongly in the preparation of the NCA3 and discussions about prospects for a sustained assessment beyond the report.

Table 1 Research goals and cross cutting capabilities for assessment

Research Goals	Foundational cross cutting research capabilities
Improve understanding of the climate system and its drivers	Integrate natural, social, engineering and other disciplinary approaches
Improve understanding of climate impacts and vulnerability	Ensure availability of observations, monitoring and infrastructure for critical data collection and analysis.
Increase understanding of adaptation pathways	Build capacity for climate assessment through training, education, and workforce development
Identify the mitigation options that reduce the risk of longer-term climate change	Enhance the development and use of scenarios
Improve decision support and integrated assessment	Promote international research and collaboration

Corell et al (2014) p 708

First, as our *understanding of the climate system* improves, assessments require that we focus on regional scale changes, possible thresholds and abrupt changes, and the interactions between anthropogenic climate change, natural variability and climate extremes. We also identified the need for development of indicators that can be used for regular reporting to policy makers and the public that are understandable and allow for attribution and anticipation of change.

Second, in terms of *understanding climate impacts and vulnerability*, the key research needs are to provide greater spatial and temporal detail in the monitoring and analysis of impacts, especially to expand beyond static snapshots of vulnerability to more dynamic analyses that reflect how climate interacts with other stresses on seasonal and inter-annual timescales to alter vulnerabilities and impacts. Other priority research needs include better understanding of the impacts of climate on particularly vulnerable groups and the interaction of multiple stresses and uncertainties. Perhaps the most important research gaps are the need for much better understanding of economic consequences of climate change, both in terms of costs, the regional and international context, and the impacts on the full range of economic sectors.

Third, there is a lack of research on *pathways to adaptation* including the identification, selection, implementation and evaluation of adaptation options. In order to identify best practices much better comparative research designs are needed that compare between different adaptation strategies and with no adaptation, and which identify the institutional and behavioral barriers to successful adaptation.

Fourth, the NCA research agenda chapter recognizes that managing climate risks, and reducing vulnerabilities, impacts and the need for adaptation, requires a priority focus on emissions reductions (*mitigation*) that reduce the risk of longer term climate change and provide synergies with adaptation at the local level. Reducing the greenhouse gas burden will reduce the severity and costs of impacts over the medium to long term. Many local decision makers are trying to manage mitigation and adaptation within single institutions and clusters of policies and need to understand the connections and trade offs between the two.

For successful *decision support* the research chapter of the NCA highlights the importance of better identifying decision-maker needs, especially under uncertainty, integrating information into risk management and developing costing and modeling tools that allow users to understand trade offs, benefits, and the implications for current action of long term scenarios.

In order to advance on these research needs, several cross-cutting capabilities are critical including support for *interdisciplinary research* that allows integration and for the infrastructure and people that provide *observations and data*. In addition the chapter identified activities that build *capacity* through education, training and workforce development that will provide the skills and people who can participate in decision-making and climate risk reduction activities. To the extent that climate risk reduction requires insight into the implications of different pathways and decisions into the future the chapter also recommended enhancing work on *scenarios*. And finally, the chapter emphasized the importance of *international cooperation* to facilitate the sharing of data collection, research responsibilities and infrastructure, to support international negotiations, and to understand the ways in which local scale impacts in the US are affected by climate changes elsewhere and global economic linkages (see below).

The NCA also includes a chapter on the design and core features of a sustained assessment – the ongoing process of understanding vulnerabilities and responses to climate variability and change that supports the adaptive management of climate risk in the United States. Several of the research needs mentioned above are especially important to a sustained assessment including capacity and network building, data collection, and the development of indicators and scenarios. A special report on sustained assessment (Buizer et al 2013) additionally recommends the development of better valuation methods, greater attention to the international context of US climate impacts, and more rigorous evaluation of assessment activities so as to support ongoing learning within the assessment community and supporting agencies.

The next section expands on several of these research gaps, illustrating their significance and some challenges in filling them in. This has required some new data analysis, theoretical framing, and synthesis of case studies that were not fully incorporated into the NCA and other recent assessments.

3 It's the economy….

The 2014 NCA report is one of many recent assessments that pay inadequate attention to key sectors of the economy, especially manufacturing and services. It includes chapters on climate impacts on natural resources such as water, forests, land and ecosystems, on urban, coastal, indigenous and rural communities, and on transport, energy and health. While these chapters include some discussion of climate impacts on infrastructure, food systems, or buildings they do not adequately address the broader effects on the economic sectors that drive national and regional productivity and employment. Emergent effects such as competition and trade offs for capital, labor and government funds and taxation are also overlooked, especially as funds are diverted to respond to climate change. This is partly due to the fact that that there is very little underlying research base to assess. For future assessments to improve on this gap, fundamental research on these sectors and connections is required. It is particularly important to look at the impact of changes in extremes.

Agriculture contributes just over 1 % to US Gross Domestic Product (GDP), whereas industry and manufacturing constitutes 20.5 % and services are 78 % (Table 1). Industry includes mining, construction, utilities, transport, and the manufacturing of durable goods (e.g., machinery, electronics, motor vehicles), food and beverages, textiles and paper, and chemicals. Each of these contributes more to GDP than the whole agricultural sector. The service sector includes tourism, information, arts and media, finance and insurance, real estate, retail, law and professions, waste management, education, health care and government. Finance and real estate, professional services, and government contributed the most to USGDP in 2012 (see Table 2).

In terms of employment, the agriculture, forestry, and fishing sector is only 1.6 % of jobs, whereas services are 79.5 % and industry and manufacturing are 13 % of employment (US Bureau of Labor Statistics 2015 www.bls.gov). Most Americans work in manufacturing (8 %), retail (10 %), professional services (12 %), education and healthcare (14 %), arts, hospitality and tourism (10 %) and government (15 %).

While some sectors of manufacturing and services are discussed in assessments and in the research literature others are almost completely overlooked and under researched, especially given their significance to the economy and employment. Tourism, transport and insurance have received the most attention; textiles, auto and electronics, retail, and IT are rarely even mentioned. Although there are hundreds of studies on climate impacts on agriculture hardly any of these track the impacts through commodity chains and the food system to assess the effects on the enormous food wholesale and retail sector.

This lack of attention to important economic and employment sectors is a research gap in terms of full accounting of climate impacts and a national and regional response to climate change, and also has serious implications for public and policy awareness of climate risks and opportunities. If research overlooks the sectors where most people work, and which drive our

Table 2 Economic sectors and the US Economy in 2012

Economic Sectors	Value added as % of GDP$_1$	Employment (thousands of jobs)	Percent employment$_2$
AGRICULTURE, FORESTRY, FISHING AND HUNTING	1.2	2122	1.6
INDUSTRY	20.5		13.00
Mining, oil and gas	2.6	800	0.60
Utilities	1.7	554	0.41
Construction	3.6	5640	3.9
Manufacturing Durable Goods	6	11918	8.2
SERVICES	78.3		79.5
Services Wholesale trade	5.9	5672	3.9
Services: Retail trade	5.7	14875	10.2
Services: Transportation and warehousing	2.9	4414	3.0
Services: Information	4.8	2677	1.8
Services: Finance, insurance, real estate, rental, and leasing	19.5	7786	5.4
Professional and business services	11.9	17930	12.3
Educational services, health care, and social assistance	8.2	20319	14.0
Arts, entertainment, recreation, accommodation, and food services	3.71	13745	9.5
Government services	13.5	2113	15

Data Source: 1. US Bureau of Economic Analysis 2015 http://bea.gov/industry/ (Accessed April 2014), 2. Bureau of Labor Statistics 2015 www.bls.gov (Accessed July 2015)

economy, then it is not surprising that many key decision makers and the public are inattentive to assessments and the risks of climate change.

Even at the regional level, agriculture and forestry are not very important to state economies. No state obtains more than 15 % of their GDP from agriculture and forestry, and only the Dakotas, Iowa, Kansas, Montana, Idaho, and Nebraska have agriculture and forestry as more than 5 % of state GDP (US Bureau of Economic Analysis 2014 http://bea.gov/iTable/index_regional.cfm).

Why have industry and services been overlooked in climate impact research and assessments? Although the IPCC did include a chapter 'Key Economic Sectors and Services' in the latest report (Arendt and Tol 2014) it focused on climate impacts on energy and water demand and supply and on transport, tourism and insurance. Only a brief section and a few references commented on the impacts on mining, manufacturing, real estate, finance, and health services. Indirect impacts on the economy were assessed through general equilibrium models with very little discussion of supply chains. The report states "For most economic sectors, the impact of climate change will be small relative to the impacts of other drivers (p 662) and "Economic activities such as agriculture, forestry, fisheries, and mining are exposed to the weather and thus vulnerable to climate change. Other economic activities, such as manufacturing and services, largely take place in controlled environments and are not really exposed to climate change" (p. 688). However the impacts on manufacturing and services from Hurricane Sandy, of cooling or pollution control costs during heat waves, and on supply chains from the Bangkok floods suggest that these sectors suffer considerable exposure to climate change (see references below). The IPCC chapter does admit, "Not all key economic sectors and services have been subject to detailed research" (p 663) and proposes more research is needed on manufacturing and services.

Both the IPCC and NCA overlook literature in business and economic geography that would provide important insights into supply chain vulnerabilities and climate impacts on key sectors (e.g., Jarmin and Miranda 2009; Leichenko and Thomas 2012; Linnenluecke at al 2011; Peck 2006; Webb et al 2000). The coastal and urban chapters of both the IPCC (Field et al. 2014 Chapter 8) and NCA (Melillo et al. 2014 Chapters 11 and 25) assessments do discuss the impacts of climate change on ports, infrastructure and energy dependent industries but the manufacturing and service sectors are given little attention beyond cities and coasts. The technical inputs to the NCA on infrastructure and urban systems and on energy supply and use provide useful overviews of what is known about impacts on energy exploration, generation and costs and on urban infrastructure, especially transportation (Wilbanks and Fernandez 2012; Wilbanks 2012). Although mention is made of supply chain links and of impacts on energy manufacturing there is little detail on key sectors such as retail, finance, or goods manufacturing except by inference from discussion of energy, water or infrastructure.

The transport and coastal chapters of the NCA do raise concerns about how climate impacts on US infrastructure may affect trade through, for example, sea level rise impacts on ports and canals or flooding of road and rail routes.

Many journals (e.g., Disasters, Natural Hazards) have articles that discuss the impacts of extreme climate and weather events on business (e.g., Marshall and Schrank 2014; Schrank et al 2013; Xiao and Nilawar 2013). Research after Hurricane Sandy is providing important insights into which sectors suffer and benefit from climate extremes in the U.S. (Leichenko et al 2014; Mantell et al 2013; Rosenzwieg and Solecki 2014; US Department of Commerce 2013). These studies show that the impacts of Hurricane Sandy were severe in the short term – with almost a billion dollars in lost tourism revenues, another billion in property damage, more than 50,000 jobs lost and declines in productivity in the utilities, chemicals, food and electronic industries – but there were some benefits for the construction sector during longer term recovery and many sectors rebounded within a year.

Floods, winds and service disruption affected more than 10,000 manufacturing facilities and in New Jersey almost 20,000 small businesses were damaged (US Department of Commerce 2013).

One reason for the lack of attention to manufacturing and services is that the government departments with responsibility for the service and manufacturing sectors are less likely to have staff with expertise relevant to environment and climate change and do not play a major role in the USGCRP.

Future assessments should take account of a broader scope of literature, especially in business and on natural disasters. There is clearly a need to encourage more research on climate impacts on manufacturing and services including case studies, detailed modeling and following secondary impact (Dell et al 2013). Disaggregating sectoral impacts provides insights into distributional effects whereby some sectors lose and others benefit from climate extremes and change – such as when a construction boom, fueled by government relief and insurance payments, rebalanced regional economic productivity following Hurricane Sandy.

4 Climate impacts in a globalized economy

Understanding how contemporary economies function is also critical to understanding how communities are affected by climate change far beyond their local region, and reveals how studies that do not pay attention to global links can misjudge local climate impacts. The NCA and many other regional climate impact studies generally do not take account of the global context for local climate impacts. For example, in most regions of the US the supply of food, energy and other goods is imported from other regions and countries, so that only looking at how climate affects local agricultural, energy and economic production provides only a partial window on how climate change and variability affect US communities. Most major companies operate within global supply and value chains, where for example, the automobile industry and major food retailers can be affected by climate events far across the world (Jira and Toffel 2013).

The best example of supply disruption is probably that of the 2011 floods in Thailand that damaged an important manufacturing center around Bangkok (Benfield 2012; Courbage and Stahel 2012). As business adopts 'just in time' supply of key parts or depends on raw materials and manufactured goods in climate vulnerable agglomerations of facilities the risk of climate disruption can increase (Smith 2013). When serious flooding inundated industrial estates around Bangkok the estimated loss of manufacturing production cost over $40 billion US with impact on more than 14,500 companies and critical supply chains for automobiles and computers (Abe and Ye 2013; Amado et al 2013). Nissan and Toyota both had to suspend production worldwide because of problems in obtaining parts from Thailand. Only ¼ of the loss was insured (Courbage and Stahel 2012) and global prices of hard disks doubled because almost half of the world's hard disks were produced in Thailand. The vulnerability of these global supply chains was increased by geographical conditions in Thailand including construction on flood plains, deforestation upstream, and problems in water governance (Haraguchi and Lall 2013).

The humanitarian NGO Oxfam has identified supply chain vulnerability as a key component of corporate responsibility in which business needs to understand and manage the impacts of climate on the small-scale producers they depend on (Thorpe and Fennell 2012). They use case studies of coffee company Starbucks and retailers Marks and Spencers and the Body Shop to show how the supply chains of these companies are vulnerable to climate change impacts on coffee, cotton and sesame oil, and the strategies that can reduce the vulnerability of source regions and producers.

There are lots of examples of the dependence of other U.S. regions on imports and global supply chains. For example, Table 3 shows key categories of imports into the US. Of total 2012 imports of $1.8 trillion dollars, almost 1/3 of the value was machines (e.g., computers), 14 % transport (e.g., cars), 25 % chemicals, minerals and metals, and about 6 % each for textiles and food. Economic vulnerability to climate shocks and change beyond the US depends on the sources of imports. In 2012 44 % of US imports were from Asia (mostly China, Japan and South Korea), 23 % from Canada and Mexico and 23 % from Europe (Table 4).

Climate impacts are also mediated by prices that are established over very large regions in a world dominated by free trade. The price of grains, oil, minerals and other goods are set within global commodity markets that can be driven by events and speculation anywhere within a global system. The spike in food prices that occurred in 2008-09 (Fig. 1) was driven by a combination of increased demand for biofuels worldwide and for meat and dairy in Asia, increased prices for energy and inputs, and climate caused declines in production of basic grains in regions as far apart as Australia and Eastern Europe (Ingram et al. 2010; Von Braun and Tadesse 2012). Speculation exacerbated volatility and the price increases prompted many farmers to put more land and inputs into production in subsequent years (Clapp and Helleiner 2012; Clapp 2014).

As an example, a freeze that damages citrus production in Latin America can benefit producers in Florida who can sell at higher prices; sour cherry production in the US Midwest interacts through markets with production in Poland. The globally connected food system is a powerful example of how even when climate has some negative effects on local agriculture, farmers can still benefit if things are worse elsewhere and they can obtain higher prices. Those studies that focus only on yields, or which do not take account of traded goods and global price signals are providing inaccurate insights into climate impacts in a globalized world. Several recent projects are starting to look further at the need for multiregional climate assessments as they relate to international markets (Barsugli et al. 2013; Winkler et al 2010; Liu et al 2013)

5 Identifying research needs through assessment

The NCA chapter on research needs was based on a review of research gaps noted in various topical and regional chapters and through specific requests to chapter authors. In my own view

Table 3 U.S. imports of key commodities

Sector	Percent of import value
Machines and instruments	34.10 %
Vehicles	14.05 %
Chemicals and Pharmaceuticals	9.30 %
Oil	5.26 %
Other Minerals	3.47 %
Metals	8.94 %
Textiles and footwear	7.13 %
Plastic and Rubber	3.75 %
Food	2.93 %
Other (paper, wood, glass, animal and veg products, art, weapons)	11.07 %

Source: MIT Atlas of Economic Complexity http://atlas.media.mit.edu/)

Table 4 Origins of US Imports

Name	Percent
China	22.11 %
Mexico	13.61 %
Canada	9.83 %
Japan	7.79 %
Germany	5.98 %
South Korea	3.49 %
United Kingdom	2.91 %
France	2.03 %
Italy	1.98 %
India	1.88 %
Brazil	1.52 %
Ireland	1.50 %
Switzerland	1.30 %
Rest of world	24.07 %

Source: MIT Atlas of Economic Complexity http://atlas.media.mit.edu/)

this did not produce the most effective or useful list of research needs because some chapters and chapter authors did not actually identify research gaps and there was overreliance on the previous experience of the Research Needs chapter authors. Such exercises also run the risk of generating laundry lists of every research gap and individual priority, because selecting the most important research needs is very challenging. Partly because of this we also relied on National Research Council (NRC) and US Global Change Research Program reports such as America's Climate Choices and the USGCRP Strategic Plan, which devoted more time and resources than the NCA to research needs identification.

Why is it important to have a research gaps chapter in assessments? How might such a chapter be developed more rigorously in sustained or future assessments?

Drawing on my own experience I can suggest several barriers and challenges in the development of research needs reports and chapters. There are also several research articles that address the assessment process – mostly the IPCC – that are helpful in considering barriers to assessments and their role in defining research needs (Anderegg et al 2010; Berkout 2010; Griggs 2014; Grundmann 2012; Hulme 2010; Janko et al 2014; Jasanoff 2010; Jones 2013: Lahsen 2013; Lemos and Morehouse 2005; Shapiro et al 2010; Stocker 2013; Tol 2011; Viner and Howarth 2014; NRC 2007)

First, almost all assessments and National Academy reports are primarily based on volunteer efforts or time taken from other responsibilities. Authors and members of committees are drawn from the scientific community and key stakeholders who almost always have full-time jobs in universities, government, the private sector, or nongovernmental organizations. While every effort is made to include the top experts in relevant fields, some key individuals may be too busy to participate. Although most individuals are not paid, the costs of bringing people together and of even a modest support staff can limit participation, especially if international expertise is required. There may also be a tendency to rely on the usual experts because of their seniority and reliability, rather than early career scholars with new ideas and more diverse perspectives. To be sure, many early career scholars will choose to devote their time to original scholarly work in order to advance their careers and younger researchers from developing

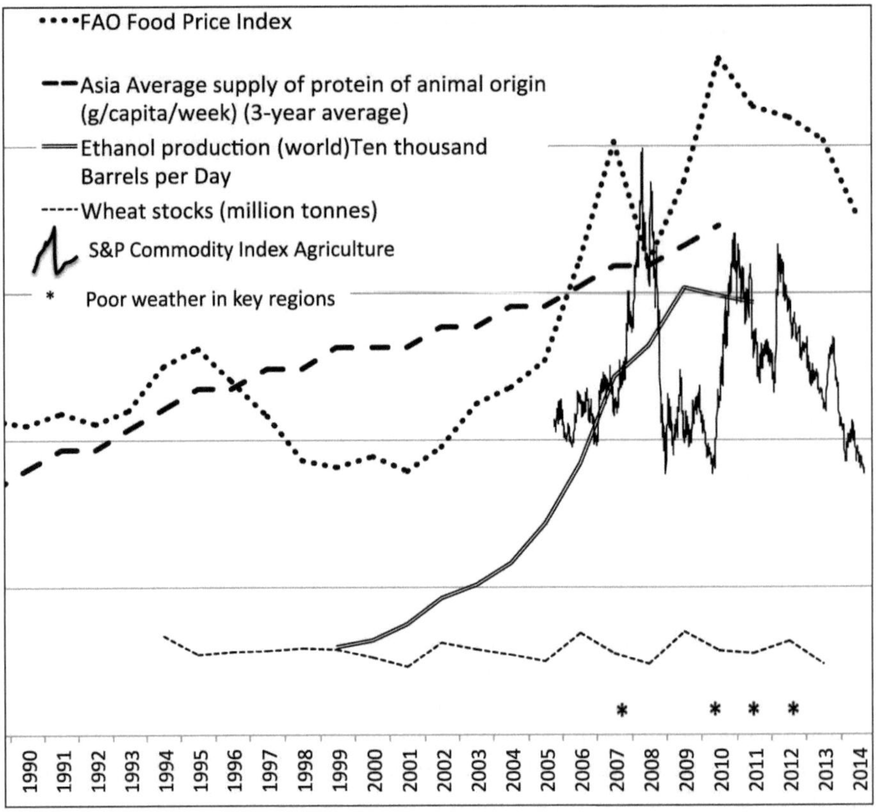

Fig. 1 The nature and selected causes of the 2007-08 food crisis. Food price index showing steep increases in 2007-08 and 2011 driven by **a**) increasing per capita consumption of meat in Asia (data from faostat.org) **b**) lower food reserves in 2008 (Data from USDA FAS) **c**) steep increase in biofuels, especially from maize (data from US Energy Information Agency) **d**) speculation in agricultural commodities as indicated by Dow Jones agricultural index with increases in 2007-08 and from 2011-2013 (data from Dow Jones S&P GCSI) **d**) bad weather in key regions. Factors not shown include increases in energy and input costs and protectionist policies

countries may face even greater financial barriers. Training in assessments and synthesis is rarely part of graduate or early career experience.

While the NCA took considerable effort to include a diverse set of scholars, my personal observation and discussions with authors suggest that some chapters and discussions were dominated by seasoned experts. Especially when it comes to research needs, younger scholars may be reading a broader range of literature.

A second consideration is the need to rely on the peer-reviewed literature – a problem mostly for the IPCC since the NCA did include publically available non-peer reviewed reports under the Data Quality Act requirements. Some assessments have been cautioned not to use reports or other publications that are not peer-reviewed even when they include information that might fill research gaps. This can be a particular problem in evaluating the impacts climate on the business community or at the local level where important case studies occur in non-peer reviewed reports or the media. Thus, information to fill the research gap is there that cannot be used. For assessments that need regional results, the constraint maybe the difficulty of publishing yet one more regional case study in a peer reviewed research literature that prizes originality. One solution is to encourage the publication of papers that review and evaluate non-peer reviewed literature in ways that represent peer review.

A third observation is the tension that sometimes emerges between identifying new research priorities and the process of funding research through government agencies. Agencies and their representatives who help write research agenda chapters may feel constrained by the politics of research funding, the time it takes to develop new funding programs and get them approved, the need to support expensive and essential long-term observations, and a reluctance to end client relationships with productive and powerful groups of scientists or institutions.

As just one example, many research agenda reports and assessments argue for more social science research but it is difficult to shift money from the natural to social sciences when most agencies employ mostly natural scientists, are more comfortable with natural science research, or fear political skepticism of social sciences (Agrawal et al 2012; Carey et al 2014; Castree et al 2014: Hackmann et a 2014; Grundmann 2012; Lahsen 2013; Stern et al 2013; Weaver et al 2014; Whitmarsh et al 2011). Calls for more social science involvement have been made many times over the years (USNRC 1992). And although NRC committees are independent, their findings can be constrained by the task they were given by the NRC or study funders.

Fourth, authors may also be reluctant to identify a large number of research gaps in case this undermines messages about what we do know. Studies suggest that when scientists to discuss uncertainties and the limits of knowledge this can reduce public and decision-maker confidence in assessments and recommendations.

Finally, climate assessments, for the most part, have approached their task beginning with the climate science and connecting through resource impacts to the rest of society. The alternative would be to begin with what matters to society, the economy, or stakeholders and then work back to understand if and how these issues are sensitive or vulnerable to climate change (Lemos et al 2012, 2014). This might result in different authors, research agendas and assessment chapters.

6 Conclusion

This paper summarizes some of the research gaps and needs that emerged from the US National Climate Assessment and takes the opportunity to reflect on the process and limitations of identifying research needs during assessment. Two research gaps are explored in greater detail – the impacts of climate change on key economic sectors and the importance of assessing impacts within a global economy. The importance of these two topics are demonstrated by discussing relevant literature that was not sufficiently included in the NCA or recent IPCC reports, and by a new analysis of economic data that demonstrates the importance of the services and manufacturing sectors and of the role of commodity chains and trade in connecting climate change and extremes elsewhere to impacts within US regions. Unless we do more research on climate impacts on manufacturing and services – which constitute most of the productivity and employment in the US and many other countries – assessments will not include the sectors where most people work or where most business and government generates growth. And unless we take account of the increasingly global connections that influence local food, energy and other commodity prices we will be unable to provide useful and accurate assessments of how climate variability and extremes affect regions, states and communities. Filling these gaps should be a high priority within sustained and future assessments if we want decision makers and the public to understand the full scope of climate change effects and if we wish to develop comprehensive responses.

Acknowledgments The author would like to thank her co-authors on Chapter 29 on Research Needs for Climate and Global Change of the National Climate Assessment including Robert Corell, Kirstin Dow, Kristie Ebi, Ken Kunkel, Linda Mearns and Jerry Melillo; Amy Glasmeier for conversations about the structure of the US Economy; and sponsors of research on drought and climate impacts including NSF EASM award 1243125, the NOAA IRAP grant to the University of Arizona and the Guggenheim Foundation.

References

Abe M, Ye L (2013) Building Resilient Supply Chains against Natural Disasters: The Cases of Japan and Thailand Global Business Review 14:567-586

Agrawal A, Lemos MC, Orlove B, Ribot J (2012) Cool heads for a hot world – Social sciences under a changing sky. Glob Environ Chang 22:329–331

Amado J-C, Adams P, Coleman H, Schuchard R (2013) PREP, Value Chain Climate Resilience. Acclimatise, Oxford

Anderegg WR, Prall JW, Harold J, Schneider SH (2010) Expert credibility in climate change. Proc Natl Acad Sci 107:12107–12109

Arent DJ and Tol RSJ (2014) Key Economic Sectors and Services. In: Field CB, V.R. Barros, D.J. Dokken, K.J. Mach, M.D. Mastrandrea, T.E. Bilir, M. Chatterjee, K.L. Ebi, Y.O. Estrada, R.C. Genova, B. Girma, E.S. Kissel, A.N. Levy, S. MacCracken, P.R. Mastrandrea, and L.L. White (eds) IPCC Climate Change 2014: Impacts, Adaptation, and Vulnerability. Part A: Global and Sectoral Aspects. Contribution of Working Group II to the Fifth Assessment Report of the Intergovernmental Panel on Climate Change. Cambridge University Press, Cambridge.

Aon Benfield (2012) 2011 Thailand Floods Event Recap Report. Aon Benfield Analytics

Barsugli JJ et al (2013) The practitioner's dilemma: how to assess the credibility of downscaled climate projections. Eos Trans AGU 94(46):424

Berkhout F (2010) Reconstructing boundaries and reason in the climate debate. Glob Environ Chang 20:565–569

Buizer, J.L., Paul Fleming, Sharon L. Hays, Kirstin Dow, Christopher B. Field, David Gustafson, Amy Luers, and Richard H. Moss, Report on Preparing the Nation for Change: Building a Sustained National Climate Assessment Process, National Climate Assessment and Development Advisory Committee, 2013.

Carey M, James LC, Fuller HA (2014) A new social contract for the IPCC. Nat Clim Change 4:1038–1039. doi: 10.1038/nclimate2442

Castree N et al (2014) Changing the intellectual climate Nature. Clim Change 4:763–768. doi:10.1038/nclimate2339

Clapp J (2014) Financialization, distance and global food politics. J Peasant Stud 1–18

Clapp J, Helleiner E (2012) Troubled futures? The global food crisis and the politics of agricultural derivatives regulation. Rev Int Polit Econ 19(2):181–207

Corell RW, Liverman D, Dow K, Ebi KL, Kunkel K, Mearns LO, Melillo J (2014) Chapter 29: Research Needs for Climate and Global Change Assessments. In: Melillo JM, Richmond TC, Yohe GW (eds) Climate Change Impacts in the United States: The Third National Climate Assessment. U.S. Global Change Research, Program, pp 707–718

Courbage C, Stahel WR eds. (2012) The Geneva Reports- Risk and Insurance Research No. 5 Extreme Events and Insurance 2011. The Geneva Association,

Dell M, Jones BF, Olken BA (2013) What do we learn from the weather? The new climate-economy literature, National Bureau of Economic Research

Field CB, V.R. Barros, D.J. Dokken, K.J. Mach, M.D. Mastrandrea, T.E. Bilir, M. Chatterjee, K.L. Ebi, Y.O. Estrada, R.C. Genova, B. Girma, E.S. Kissel, A.N. Levy, S. MacCracken, P.R. Mastrandrea, and L.L. White (eds) (2014) IPCC Climate Change 2014: Impacts, Adaptation, and Vulnerability. Part A: Global and Sectoral Aspects. Contribution of Working Group II to the Fifth Assessment Report of the Intergovernmental Panel on Climate Change. Cambridge

Global Change Research Program US (2012) The National Global Change Research Plan 2012-2021. A Strategic Plan for the US Global Change Research Program. US Global Change Research Program, Washington DC

Griggs D (2014) Climate policy: Streamline IPCC reports. Nature 508:171–173

Grundmann R (2012) The legacy of climategate: revitalizing or undermining climate science and policy? Wiley Interdiscip Rev Clim Chang 3:281–288

Hackmann H, Moser SC, Clair ALS (2014) The social heart of global environmental change. Nat Clim Change 4:653–655

Hall JA, Blair M, Buizer JL, Gustafson DI, Holland B, Moser SC, Waple AM (2014) Chapter 30: Sustained Assessment: A New Vision for Future U.S. Assessments. Climate Change Impacts in the United States: The Third National Climate Assessment, U.S. Global Change Research Program. In: Melillo JM, Richmond TC, Yohe GW (eds) Climate Change Impacts in the United States: The Third National Climate Assessment. U.S. Global Change Research, Program, pp 719–726

Haraguchi M, Lall U (2013) Flood Risks and Impacts: Future Research Questions and Implication to Private Investment Decision Making for Supply Chain Networks. UNISDR, Geneva

Hulme M (2010) Problems with making and governing global kinds of knowledge. Global Environ Chang 20: 558–564. doi:10.1016/j.gloenvcha.2010.07.005

Ingram JSI, Ericksen P, Liverman DM (2010) Food security and global environmental change. Earthscan, London

Jankó F, Móricz N, Vancsó JP (2014) Reviewing the climate change reviewers: Exploring controversy through report references and citations. Geoforum 56:17–34

Jarmin RS, Miranda J (2009) The impact of Hurricanes Katrina, Rita and Wilma on business establishments. J Bus Valuation Econ Loss Analys 4(2)

Jasanoff S (2010) Testing time for climate science. Science 328:695–696

Jira C, Toffel MW (2013) Engaging supply chains in climate change. Manuf Serv Oper Manag 15(4):559–577

Jones N (2013) Climate assessments: 25 years of the IPCC. Nature 501:298–299

Lahsen M (2013) Climategate: the role of the social sciences. Clim Chang 119:547–558

Leichenko RM, Thomas A (2012) Coastal cities and regions in a changing climate: economic impacts, risks and vulnerabilities. Geography Compass 6:327–339

Leichenko R, McDermott M, Bezborodko E, Brady M, Namendorf E (2014) Economic Vulnerability to Climate Change in Coastal New Jersey: A Stakeholder-Based Assessment. J Extreme Events 01(01):1450003. doi: 10.1142/S2345737614500031

Lemos MC, Morehouse BJ (2005) The co-production of science and policy in integrated climate assessments. Glob Environ Chang 15:57–68

Lemos MC, Kirchhoff CJ, Ramprasand V (2012) Narrowing the climate information usability gap. Nat Clim Chang 2:789–794

Lemos MC, Kirchhoff CJ, Kalafatis SE, Scavia D, Rood RB (2014) Moving Climate Information off the Shelf: Boundary Chains and the Role of RISAs as Adaptive Organizations. Weather, Clim, Soc 6:273–285

Linnenluecke MK, Stathakis A, Griffiths A (2011) Firm relocation as adaptive response to climate change and weather extremes. Glob Environ Chang 21(1):123–133

Liu J, Hull V, Batistella M, DeFries R, Dietz T, Feng F, Thomas W, Hertel RC, Izaurralde EF, Lambin SL, Luiz A, Martinelli WMC, Emilio F, Moran RN, Ouyang Z, Karen R, Polenske AR, de Miranda G, Rocha CA, Simmons PH, Verburg PV, Zhang F, Zhu C (2013) Framing Sustainability in a Telecoupled World. Ecol Soc 18:26

Mantell N.H., Seneca JJ, Lahr ML, Irving W (2013) The economic and fiscal impacts of Hurricane Sandy in New Jersey Rutgers Regional Report vol Issue Paper 34. Rutgers University

Marshall M, Schrank H (2014) Small business disaster recovery: a research framework. Nat Hazards 72:597–616. doi:10.1007/s11069-013-1025-z

Melillo J, Richmond T. and Yohe G (2014) Climate Change Impacts in the United States: The Third National Climate Assessment. US Glob Chang Res Program

National Research Council (1992) Global Environmental Change: Understanding the Human Dimensions. National Academy Press, Washington, DC

National Research Council (2007) Analysis of Global Change Assessments: Lessons Learned. National Academy Press, Washington, DC

National Research Council (NRC) Committee on America's Climate Choices (2010) Advancing the Science of Climate Change: America's Climate Choices. National Academies Press

Peck H (2006) Reconciling supply chain vulnerability, risk and supply chain management. Int J Logist: Res Appl 9(2):127–142

Rosenzweig C, Solecki W (2014) Hurricane Sandy and adaptation pathways in New York: Lessons from a first-responder city Global Environmental Change (in press)

Schrank H, Marshall M, Hall-Phillips A, Wiatt R, Jones N (2013) Small-business demise and recovery after Katrina: rate of survival and demise. Nat Hazards 65:2353–2374. doi:10.1007/s11069-012-0480-2

Shapiro HT et al. (2010) Climate change assessments: Review of the processes and procedures of the IPCC InterAcademy Council, Amsterdam

Smith MH (2013) A Fundamental Approach to Facilities Location Risk Assessment and Its Relevance to Supply Chain Network Design: Using the Thai Floods of 2011 as an Example International Journal of Trade. Econ Finance 4(5):300–3013

Stern PC, Ebi KL, Leichenko R, Olson RS, Steinbruner JD, Lempert R (2013) Managing risk with climate vulnerability science. Nat Clim Chang 3:607–609. doi:10.1038/nclimate1929

Stocker TF (2013) Adapting the assessments. Nat Geosci 6:7–8

Thorpe J, Fennell S (2012) Climate Change Risks and Supply Chain Responsibility. Oxfam International, Oxford UK

Tol RS (2011) Regulating knowledge monopolies: the case of the IPCC. Clim Chang 108:827–839

US Bureau of Economic Analysis (2014) GDP by state and sector. http://bea.gov/iTable/index_regional.cfm

US Bureau of Economic Analysis (2015) Industry and GDP. http://bea.gov/industry/

US Bureau of Labor Statistics (2015) Employment by major sector. http://www.bls.gov/emp/ep_table_201.htm

US Department of Commerce (2013) Economic Impact of Hurricane Sandy. U.S. Department of Commerce

USGCRP (Global Change Research Program). 1990. US Global Change Research Act (see http://www.globalchange.gov/about/legal-mandate)

Viner D, Howarth C (2014) Practitioners' work and evidence in IPCC reports. Nat Clim Change 4:848–850

Von Braun J, Tadesse G (2012) Global food price volatility and spikes: an overview of costs, causes, and solutions. ZEF Discussion Papers on Development Policy

Weaver C et al (2014) From global change science to action with social sciences. Nat Clim Chang 4:656–659

Webb G, Tierney K, Dahlhamer J (2000) Businesses and Disasters: Empirical Patterns and Unanswered Questions. Nat Hazards Rev 1:83–90. doi:10.1061/(ASCE)1527-6988(2000)1:2(83)

Whitmarsh L, O'Neill S, Lorenzoni I (2011) Climate change or social change? Debate within, amongst, and beyond disciplines. Environ Planning A 43:258–261

Wilbanks T and Fernandez S (2012) Climate Change Energy Supply and Use: Technical Report for the U.S. Department of Energy in Support of the National Climate Assessment. Department of Energy, Oak Ridge. Wilbanks T (2012) Climate Change and Infrastructure, Urban Systems and Vulnerability: Technical Report for the U.S. Department of Energy in Support of the National Climate Assessment. Department of Energy, Oak Ridge.

Winkler JA, Thornsbury S, Artavio M, Chmielewski F-M, Kirschke D, Lee S, Liszewska M, Loveridge S, Tan P-N, Zhong S, Andresen JA, Black JR, Kurlus R, Nizalov D, Olynk N, Ustrnul Z, Zavalloni C, Bisanz JM, Bujdosó G, Fusina L, Henniges Y, Hilsendegen P, Lar K, Malarzewski L, Moeller T, Murmylo R, Niedzwiedz T, Nizalova O, Prawiranata H, Rothwell N, van Ravensway J, von Witzke H, Woods M (2010) Multi-regional climate change assessments for international market systems with long-term investments: A conceptual framework. Clim Chang 103:445–470

Xiao Y, Nilawar U (2013) Winners and losers: analysing post-disaster spatial economic demand shift. Disasters 37:646–668. doi:10.1111/disa.12025

Climatic Change (2016) 135:187–201
DOI 10.1007/s10584-015-1530-z

Aspirations and common tensions: larger lessons from the third US national climate assessment

Susanne C. Moser[1,6] **· Jerry M. Melillo**[2] **·**
Katharine L. Jacobs[3,7] **· Richard H. Moss**[4] **·**
James L. Buizer[5,8]

Received: 1 June 2015 / Accepted: 6 October 2015 / Published online: 21 October 2015

Abstract The Third US National Climate Assessment (NCA3) was produced by experts in response to the US Global Change Research Act of 1990. Based on lessons learned from previous domestic and international assessments, the NCA3 was designed to speak to a broad public and inform the concerns of policy- and decision-makers at different scales. The NCA3 was also intended to be the first step in an ongoing assessment process that would build the nation's capacity to respond to climate change. This concluding paper draws larger lessons from the insights gained throughout the assessment process that are of significance to future US and international assessment designers. We bring attention to process and products delivered, communication and engagement efforts, and how they contributed to the sustained assessment. Based on areas where expectations were exceeded or not fully met, we address four common tensions that all assessment designers must confront and manage: between (1)

All of the authors of this paper played leadership roles in the NCA3, and are therefore not unbiased observers.

This article is part of a special issue on "The National Climate Assessment: Innovations in Science and Engagement" edited by Katharine Jacobs, Susanne Moser, and James Buizer.

✉ Susanne C. Moser
promundi@susannemoser.com

1 Susanne Moser Research & Consulting, Stanford University, Santa Cruz, CA, USA

2 Marine Biological Laboratory, Woods Hole, MA, USA

3 Center for Climate Adaptation Science and Solutions, University of Arizona, Tucson, AZ, USA

4 Pacific Northwest National Laboratory, Joint Global Change Research Institute, College Park, MD, USA

5 Climate Adaptation and Development, Institute of the Environment, University of Arizona, Tucson, AZ, USA

6 Woods Institute for the Environment, Stanford University, Palo Alto, CA, USA

7 Department of Soil, Water and Environmental Science, University of Arizona, Tucson, AZ, USA

8 School of Natural Resources and Environment, University of Arizona, Tucson, AZ, USA

core assessment ingredients (knowledge base, institutional set-up, principled process, and the people involved), (2) national scope and subnational adaptive management information needs, (3) scope, complexity, and manageability, and (4) deliberate evaluation and ongoing learning approaches. Managing these tensions, amidst the social and political contexts in which assessments are conducted, is critical to ensure that assessments are feasible and productive, while its outcomes are perceived as credible, salient, and legitimate.

1 Introduction

The Third US National Climate Assessment (NCA3) was produced in response to requirements of the US Global Change Research Act of 1990. While now a government document, the NCA3 report is also "owned" by its 240 plus authors and others who contributed to it (Jacobs and Buizer 2015 this issue; Cloyd et al., Submitted for publication in this special issue). Based on lessons learned from previous domestic and international assessments (NRC 2007), the NCA3 was strategically designed to speak to a broad public and address the concerns of policy- and decision-makers at various levels of government. Beyond these objectives, the NCA3 was intended to be the first step in an ongoing assessment process that would build the nation's capacity to respond to the challenges increasingly posed by climate change (Buizer et al. 2013). Whether or not these intentions and hopes will be realized will be determined by actions over the coming years and through independent evaluations of the successes and failures of the NCA3. Meanwhile, this Special Issue provides detailed inside views on how the assessment was conducted, and focuses attention on areas where significant progress has been made and others where the assessment's designers' own expectations were not fully met. Although its focus is US-specific and its design is a consequence of the political and economic circumstances that prevailed during its development, the NCA3 is a prominent example of common challenges and tensions that all assessment designers and implementers may face (e.g., NRC 2007; Clark et al. 2006). It also serves as an example of how to address them. In this final paper we identify cross-cutting lessons and synthesize some of the most salient advice from the authors across the Special Issue.

In doing so, we clearly recognize that all assessments must respond to specific legal requirements, charters, or policy questions, and that institutional and fiscal conditions will be significant factors in shaping the "art of the possible." At the same time, the design of an assessment is always motivated (or burdened) by multiple – and sometimes divergent – expectations of what it is intended to accomplish (NRC 2007). The federal advisory committee for the NCA3, the National Climate Assessment Development and Advisory Committee (NCADAC), for example, explicitly decided to go beyond the minimum task of summarizing and assessing the state-of-knowledge on climate change in order to better meet the needs of society. In the absence of a fully developed US National Climate Service (Miles et al. 2006), some NCA3 participants hoped that the assessment would fill some of the information needed in adaptive management and planning. The NCADAC also chose to build "knowledge networks" involving scientists and decision-makers who could share expertise and together identify the most critical issues for the assessment. This was intended to enhance the relevance of the NCA3, but also to build adaptive capacity across regions and sectors of the US.

The high-level lessons detailed below are selected on the basis of their wider relevance, intending to inform and inspire future assessment experimentation, both nationally and internationally.

2 NCA3's significance to assessment designers

Given ongoing rethinking of major international assessments, such as those conducted by the Intergovernmental Panel on Climate Change, the Intergovernmental Platform on Biodiversity and Ecosystem Services, the OECD and others, there are a number of reasons why the lessons learned from the NCA3 might interest policy-makers and assessment leaders internationally.

First, national climate assessments are an essential foundation for adaptation and mitigation work undertaken within countries, but also serve as a foundation for international policy discussions. US federal agencies (e.g., Department of Commerce, USAID, Department of State) use such assessments to support international policy development which affect government activities beyond US borders. For example, NCA3 served (even while still in draft form) as the foundation for the most significant Executive Branch commitment to international and domestic climate action in US history. It has been explicitly cited as the scientific basis for dozens of resilience-related government efforts, including the President's Climate Action Plan (Executive Office of the President 2013).

Beyond its direct relevance to policy, NCA3 is of interest to international audiences because of the innovative approaches and strategies employed in its design and conduct. Among these innovations were its risk-management framing, the treatment of uncertainty (Moss, Submitted for publication in this special issue), experimentation with the role of assessments in building capacity for adaptation (Jacobs and Buizer 2015 this issue), the process documentation provided in a series of methodology workshops and associated assessment methodology reports, the innovative electronic delivery of the final product (Waple et al., Submitted for publication in this special issue), and the development of indicators of change (Kenney et al., Submitted for publication in this special issue). We do not discuss these particular innovations here again (see cited papers), but focus instead on the larger, cross-cutting lessons learned in the NCA3 below.

Due to the integrated nature of human-environmental systems, the risks and opportunities associated with climate change cascade across sectoral and international boundaries. This makes a clear-eyed assessment of risks in any country of mutual interest to those managing emissions and minimizing risks across the globe. Even climate drivers that cause direct local impacts in certain places (e.g., sea-level rise and related hazards affecting specific areas of the coast) can result in indirect ripple effects (e.g., via the insurance and reinsurance industry or international trade) in other regions, sectors, and nations (Moser and Davidson 2015 this issue). Negative impacts on Indigenous lands and cultures are of global significance (Maldonado et al., Submitted for publication in this special issue). Similarly, economic impacts due to changes in ecosystem services (Grimm et al., Submitted for publication in this special issue), agricultural productivity, and so on occurring in the US will indirectly affect other countries. The converse is also true. In a globalized economy, nations are inextricably linked in myriad ways, and these linkages can be explored through assessment processes (e.g., Moser and Finzi Hart 2015). Although the NCA3 does not fully articulate the international implications of climate change on the US (Liverman 2015 this issue), there are multiple instances where these issues are discussed in the NCA3. More fully evaluating the issue of international interdependencies is an unfilled priority that needs to be addressed in future US assessments (Buizer et al. 2013), and is a good example of the transferable research and assessment needs that every nation faces.

Given the important role assessment play in documenting where impacts are manifesting and whether mitigation and adaptation efforts are sufficient to prevent and/or limit harm, a fuller appreciation of the interaction between assessments and climate responses (e.g., Howden and Jacobs, Submitted for publication in this special issue) and the need for evaluating decision-support needs and capacities (e.g., Moss, Submitted for publication in this special issue) is relevant well beyond the US experience.

Furthermore, the US investment in climate science is the largest of any country globally (see annual research budgets in the "Our Changing Planet" publications at www.globalchange.gov). Periodic efforts to integrate, analyze and evaluate the outcomes of this investment also serve as a critical contribution to climate knowledge for the global community. The integration of multiple kinds of knowledge – ranging from physical climate science, to social science, to tacit knowledge of resource managers – enhances the relevance of the findings of the US Global Change Research Program (USGCRP), serves as a contribution to knowledge well beyond the US, and is of special interest to scientists and policy-makers worldwide.

Finally, an important innovation is the unprecedented degree of public engagement and transparency that permeated the NCA3 process (Cloyd et al., Submitted for publication in this special issue). There was substantial perceived risk to the process itself associated with the decision to maximize engagement of many regional and sectoral experts who had never before been engaged in climate assessment processes. Though this risk was a source of concern to many and was used as a rationale for cutting back on public outreach and engagement on occasion, in retrospect, these efforts proved very successful and may actually have served as a kind of shield against unwarranted criticism for the process and the outcomes. This is an important lesson for others as they move forward in assessing climate risks and opportunities.

3 Process, products, engagement and communications strategy, and the sustained assessment

The charge given to the NCADAC, which oversaw the NCA3, had two parts. The first part, as set out in the NCADAC charter, was "to synthesize and summarize the science and information pertaining to current and future impacts of climate change upon the United States." The second was "to provide advice and recommendations toward the development of an ongoing, sustainable national assessment of global change impacts and adaptation and mitigation strategies for the Nation." This two-part charge guided the work of everyone involved in the design and production of NCA3 and framed the recommendations submitted to the USGCRP regarding a sustained assessment process (Buizer et al. 2013). Here, we reflect on three core aspects of NCA3 – its process, its products, and its engagement and communication strategies – and how these aspects relate to the sustained assessment, as a first way to synthesize insights and lessons learned from the assessment.

3.1 Process

Participants in the NCA3 generally agree that it was successful in building an actively engaged assessment community while also promoting engagement of stakeholders and informing decision-making. From the perspective of the authors of this article, important process successes included the internal assessment management and staffing, the consensus decision process established by the NCADAC, the structure and leadership provided by the Executive Secretariat (established to help manage the process given the 60-member NCADAC), the overall transparency of the process, the active engagement of federal agencies (in some instances significantly more than in previous assessments), the broad participation of both knowledge users and producers, and the fostering of strong regional climate-assessment nodes (Cloyd et al., Submitted for publication in this special issue).

Beyond the explicit focus on including new faces and multiple perspectives in leadership and author teams, the expanded role of the federal agencies in this process[1] not only facilitated financial support for the assessment, but it also fostered participation of talented federal scientists with expertise in how to access and interpret information (e.g., large government-collected datasets) relevant to the assessment (Waple et al., Submitted for publication in this special issue). The NCA3 also successfully built interdisciplinary author teams (including social scientists) and asked process experts to help facilitate effective interaction among diverse participants.

Based on discussions with NCA3 authors and stakeholders, the broad participation fostered engagement and dialogue. In a number of cases, this led to the co-production of new knowledge (e.g., in the coastal chapter; Moser and Davidson 2015 this issue). The strong regional assessment teams supported regional and sub-regional analyses of climate impacts and geographically relevant responses, each producing peer-reviewed and subsequently published "foundational documents." These teams also provided a structure from which to engage decision-makers and inform stakeholders potentially interested in climate changes, impacts, and responses.

Not all aspects of the assessment process were unqualified successes, however. Areas in which the process fell short included inconsistent deployment of the risk-based framing and the guidance on documenting degrees of certainty in the findings (Moss, Submitted for publication in this special issue). Looking to the future, the framing approach and instructions about characterizing degrees of certainty must be communicated to authors early and clearly (as noted in NRC 2007), and applied consistently across all outputs produced in the sustained assessment process. In addition, while external experts had recommended including risk communication and uncertainty assessment experts on every team, the realities of timing, author selection criteria, and resources precluded this. Similarly, formal evaluation of assessment and decision-support processes was not as well integrated into the NCA3 as it could have been. Improving the dialog about decision-maker needs and useful products at all stages of the assessment process would strengthen future assessments efforts (Moss, Submitted for publication in this special issue). The methodology workshops and specific training workshops for authors before assessment activities commence could be used for such capacity building.

3.2 Products

From the perspective of the authors of this paper, NCA3 products can be judged successful along various dimensions. They include technical reports that served as inputs to the regional and most of the sectoral chapters of the synthesis report, the highly integrated and very clearly written highlights document, and an exceptional web site. The technical reports not only helped to provide a solid foundation for the public review draft of the 2014 quadrennial report, they also contributed to achieving a major objective of the assessment – engagement of a broad set of experts and stakeholders in the process. The highlights document captured the essence of the much longer synthesis report in a way that proved very accessible to the press and played an important role in the extensive coverage of the NCA3 in newspapers (over 10,000 articles published within a year of the release of the NCA3), on radio and TV, and on new and social media. Web statistics suggest that regional chapters are of particular value to readers – most likely a reflection of people's interest in what happens in the places where they live. The

[1] Each federal agency was represented by an *ex officio* member on the NCADAC.

overview document and the NCA3 web site have both proven to be excellent teaching tools, and are being used widely to educate and inform the public about climate change, its impacts, and possible adaptation and mitigation responses (Cloyd et al., Submitted for publication in this special issue). Another successful product-related innovation was the initiation of a national indicator system, which is intended to be regularly updated to help monitor changes over time in a more consistent way (Kenney et al., Submitted for publication in this special issue) .

The biggest product-related challenge came when the deadline for completion was looming. Complex parallel processing occurred at this time and involved the simultaneous production of final chapters and multiple synthesis materials, as well as the creation and deployment of the website; it proved challenging to do all of this while maintaining consistency across all of the products. However, these challenges were mostly overcome through the work of dedicated USGCRP and Technical Support Unit staff.

3.3 Communication and engagement

Internal and external communication about NCA processes and products was highly effective, thanks in part to a dedicated working group of the NCADAC that worked with USGCRP staff to develop relevant strategies early in the NCA3 effort. The assessment distinguished communication (two-way exchange about the assessment) from engagement (forms and opportunities for participation in the assessment process).

Regarding the former, the staff and technical editors' efforts to produce written documents (and associated graphics) in an accessible style and consistent voice, and the online delivery with easy links to social media, were critical elements in successful communication (Cloyd et al., Submitted for publication in this special issue). Regarding the latter, the creation of the NCA network (NCAnet) was perhaps the single most innovative aspect of the NCA3. NCAnet served as one of the principal venues for engagement of outsiders with the NCA process, and continues as a "network of networks" that engages people across the US. The more than 150 NCAnet partner organizations have extended the NCA process and products to a broader audience than could have been reached otherwise. They have developed assessment-related capacities and products, hosted workshops, town halls and media events, collected and synthesized data, and produced technical and scientific information relevant to current and future NCA reports. In addition, NCAnet partners have disseminated report findings to various audiences and produced secondary products related to the NCA and NCA findings. For example, the National Council for Science Education supports an education affinity group within NCAnet that has produced curricula, webinars and other training sessions based on NCA3 content. Further, NCAnet partners helped produce – for the first time – a summary of NCA3 findings in Spanish for the 50+ million Americans who speak Spanish in their homes daily.

A major challenge in communication involved the management of expectations of the public after the public review period, but before the release of the completed report. Thirteen months elapsed between the close of the public review period and issuing the final NCA3. During this time, four important activities occurred: (1) revisions of the public review draft in response to comments; (2) evaluation by review editors of the degree to which the revisions were responsive to the public review comments; (3) several rounds of technical review of the revised NCA3 drafts by scientists at USGCRP agencies and at the Office of Science Technology Policy; and (4) signoff by report authors, and subsequently the NCADAC, on the final draft of the report. The public was interested in following progress of the process and could do so in a general way, but did not get to see each iteration. Carefully managing the need for transparency and legitimacy and the need for efficient completion of the full report without undermining its credibility or its salience was a challenge in the NCA3 and is an issue for all large-scale assessments.

3.4 Sustained assessment process

The NCA3's overarching goal of establishing a highly credible, ongoing assessment process was linked to three sub-goals: building a foundation of engagement with assessment partners, creating easily understandable and accessible products that can be updated on an ongoing basis, and ensuring transparency and testing effective assessment processes. Details of how to best implement this set of goals are still evolving within the USGCRP and are guided, in part, by a special report produced by the NCADAC, *Preparing the Nation for Change: Building a Sustained National Climate Assessment Process* (Buizer et al. 2013). Some recommendations have already moved from idea to action (Buizer et al. 2015 this issue). For example, targeted topical assessments were suggested that meet user- or expert-identified information needs and can serve as inputs into subsequent quadrennial reports. The first of these targeted reports, entitled *The Impacts of Climate Change on Human Health in the United States: A Scientific Assessment,* is being completed at the time of this writing. A USGCRP Indicators Initiative (Kenney et al., Submitted for publication in this special issue) is now in a pilot phase. Workshops have been held on the development of land use/land cover and population scenarios for use at sub-national scales. Additionally, the preparation of usable climate scenarios is being discussed.

Other topics and recommendations appear to be lagging. Of greatest concern is that the focus of activity remains almost exclusively on the production of reports coordinated through federal agencies. The special report's recommendations to move toward a wider range of NCA products such as data sets, scenario planning methods, tools for vulnerability assessments, maps, and others, and to make a more concerted effort at addressing the international dimension have not yet been realized. The recommendation to encourage a distributed assessment approach, allowing for a series of self-motivated assessment processes organized by municipalities, sectoral interest groups, universities, NGOs and other interested parties does not seem to have progressed. The recommendation to appoint a smaller advisory committee with expertise appropriate for assisting with the transition to a sustained process is still in early stages at the time of this writing. Steps to evaluate and improve approaches for characterizing uncertainty and communicating confidence do not appear to have been taken. The apparent lack of progress in some of these areas is of concern since the clock is ticking toward the 2018 quadrennial report.

4 Common tensions requiring management in effective assessments

In the course of NCA3, we encountered a number of tensions related to diverging expectations about process, products and outcomes, information needs, available resources, size and complexity of the effort, and other issues. These tensions emerged in discussions about how to build a workable structure for an effective assessment, how to choose topics at scales that are most useful given the wide range of possible approaches, how to ensure efficient outcomes and processes that are consistent with agreed-upon goals for an assessment, how to agree on priorities across multiple expectations of what assessments ought to achieve, and how to critically, but constructively, learn from ongoing assessment efforts so as to continually improve them. These are challenges all assessment designers and leaders face.

Perspective gained on these topics during the NCA3 may be useful to other assessment processes and are offered here as another way to extract larger lessons from the contributions in this Special Issue. The tensions and trade-offs are sometimes more perceived than real, but difficult choices clearly have to be made in any assessment. Below we describe issues and choices made in the NCA3. The effectiveness of future assessments may hinge on how these tensions are managed given place- and time-specific circumstances.

4.1 Tension among core assessment ingredients

Previous studies focused on the tension between "product" and "process" in shaping assessments (e.g., Morgan et al. 2005; Clark et al. 2006; NRC 2007). This framing brought central attention to the outputs of an assessment and how those co-determine the most important qualities that assessment participants and observers aspire to: salience, credibility and legitimacy (Cash et al. 2003). Together, these qualities affect the overall effectiveness of an assessment in terms of informing policy and decision-making. What the product–process dichotomy hides are the principal ingredients that enable particular products and processes to be delivered. In the interest of informing future assessment designs, however, it is useful to focus on those essential ingredients, and thus deepen the understanding of qualities that render some assessments more effective than others. Such a deeper look brings attention to the *inside* workings of assessments, i.e., to the underpinnings that allow products and processes (Section 3) to be delivered.

Based on our experience from multiple assessment processes, four ingredients are critical to success: (i) the information, data, scientific studies, situational knowledge (of real world problems, geography, systems or decision contexts), and other kinds of traditional knowledge that form the knowledge base of the assessment; (ii) the institutions and resources to conduct an assessment (reflecting legal support, legacies of other processes and reports, financial and staff resources, political will and commitment); (iii) principles and approaches that guide participants; and (iv) the people/human capital required to carry it out (Fig. 1). We view each of these as necessary but – by themselves – insufficient ingredients that enable an assessment to deliver products and processes that eventually are perceived as credible, salient, and legitimate. While there is some degree of mutual compensation across these four essential ingredients (dedicated people can make up, to some extent, what money could buy if it were available; a principled process can compensate for some institutional inadequacies), assessment designers cannot ignore any of these four essential ingredients unless they are willing to risk the success of the whole operation. These core ingredients, to whatever extent available, must be aligned so that all assessment participants and sponsors come – through the course of ongoing communication, dialogue and negotiation – to hold a shared understanding of and agree to:

- the mandate and expectations about outcomes and products, including an understanding about who the likely assessment users will be and how they might use its products;
- the rules of engagement, standards of procedure, and legal/institutional requirements;
- the identification and roles of different players, including leaders, technical experts, intermediaries to help with interpretation and communication, editors, reviewers, graphic artists and policy- or decision-making communities with the capacity to engage in the process and apply the knowledge generated; and
- the logistical constraints (timing, fiscal) that affect what can be done.

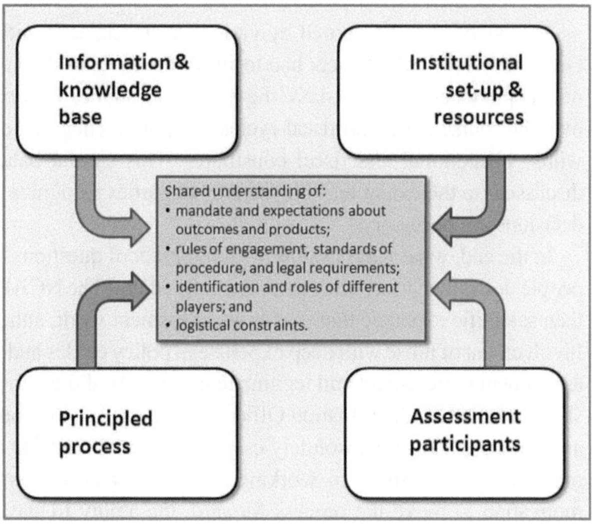

Fig. 1 The art of the possible: Careful alignment of core assessment ingredients ensures an effective assessment operation. Source: The authors

How were the tensions among these four core ingredients handled in NCA3? First, much attention was paid to the knowledge foundation. Clearly, every decision made about the information base affects its perceived salience, credibility, and legitimacy. For example, the quality and accessibility of underlying data to foster transparency and utility of the final product were important foci. Similarly, a widely recognized strength of NCA3 was the organization and presentation of observations of climate changes at various spatial scales, from global, to national, to regional. Rigorous scientific procedures and criteria were used to attribute specific impacts to climate change vs. other drivers of change at each of these scales. Another example is the openness with which, in the early days of the NCA3, a number of important questions were discussed and resolved between the authors and the NCADAC, adding to its ultimate credibility and salience, e.g.,: Which climate change scenarios should be used so as to recognize the state of the science and be compatible with IPCC, yet also provide information at the right scale and resolution for every region of the US? To what extent should the state of physical climate science dictate which ecological and social impacts science can be used? How should non-peer-reviewed information be handled, i.e., what is a principled and defensible approach to quality assurance? How do we track, inventory, store, and make available and accessible all the information that went into the production of the NCA3? And finally, what delivery mechanisms and product formats are most useful to citizens and decision-makers?

Clarity about rules and procedures as well as about the roles and responsibilities of different assessment participants – the second and third core ingredients – was critically important. For example, extensive effort was put into designing an assessment process that built on past experience with other assessments, took advantage of the best available methodological approaches (process and methods workshops), but also was feasible in the current political and economic context within the US. Assessment leaders, sponsoring agencies and the NCADAC deliberated on: How much engagement is useful, desirable, necessary, affordable, and feasible? What are the best engagement and communication approaches? How can federal agency needs and capacities be balanced with external stakeholder desires and capacities? How much can be asked of volunteers? How can ongoing evaluation and learning be ensured without endangering timely progress and support from all involved?

The NCA3 was sponsored by various agencies, each with particular rules and procedures. Given these, federal partners had to fulfill the letter of the law, while also experimenting with new approaches to better serve the American public. This raised continuous questions and – over the course of several fiscal cycles – required renegotiation of what is prudent and possible within institutional and fiscal constraints. This careful balancing act needs to be explicitly discussed (to the extent feasible) so that all parties recognize why certain process management decisions are necessary.

In the end, what was possible within this set of questions and constraints came down to the people doing the work – the fourth core ingredient. The NCADAC understood that it is far more than scientific expertise that makes an assessment work, although that is crucial for credibility. Involvement of those with deep expertise in policy circles and management could help make the assessment more salient and legitimate in the eyes of the public and policy-makers. Leadership skills in the NCA Coordination Office, the NCADAC, its working groups, within author teams, and among staff were absolutely essential: commitment (by everyone, paid or unpaid), determination and creativity in working around obstacles, openness to input, decisiveness, and motivation to move the process forward, the ability to stay organized and engaged with 30 chapter teams moving forward simultaneously, and so much more. These "people skills" were decisive in delivering a salient, timely product. At the same time, considerable effort went into building social capital among all involved: within the NCA Coordination Office, within the NCADAC and its working groups, within author teams, and among the extended "NCA family" (e.g., NCAnet). With a very tight budget and a largely voluntary effort, the importance of investments in people and relationships cannot be overestimated.

4.2 Tension between national scope and Sub-national adaptive management information needs

As the NCA seeks to become increasingly relevant to adaptation and emissions reduction efforts, the tension between providing information at national and more local scales will increase. Users involved in adaptive management and planning expect to receive information tailored to their location, decision or question, but available models and data sources are often not yet up to this task. Information at the national or regional scale can be helpful in framing challenges or providing a broad context, but may be too coarse to inform certain decisions. In addition, decision support for planning or managing specific adaptation and mitigation efforts will necessarily entail more user support and engagement, leaning towards "climate services." It seems unlikely that a centralized *national* assessment will ever have sufficient resources to provide a comprehensive climate services portfolio to meet *local* needs. Thus assessments will need to balance user expectations about fulfilling information needs with what is technologically possible and scientifically defensible. Galford et al. (Submitted for publication in this special issue), explore the lessons learned in a state-scale version of a climate assessment, providing insights useful in this regard.

One possible extension of traditional assessments, however, and thus one way to better translate between what is scientifically defensible and practically desirable, is for the sustained assessment process to not just focus on comprehensive national reports and periodic topically-targeted reports, but also on critically assessing the use and value for specific purposes of commonly available data, methods, visualizations, and other tools and resources. An assessment cannot provide decision support systems tailored to individual decision-making environments and situations, but it can add value by producing credible and useful outputs with wide applicability. It can also help assessment designers and users to better understand and reflect on the respective roles and relationships of national and sub-national assessments, including a clearer division of labor among public and private entities providing climate services. In addition, it can synthesize knowledge about effective decision support, evaluate different types of systems, and provide good practice guidelines (Moss, Submitted for publication in this special issue).

Another approach that emphasizes the role of assessments in promoting dialog between decision-makers and the research community is to explore information needs and available products to meet those needs. Users often request specific products or methods (such as downscaling), when in fact, upon reflection with decision-support experts, it becomes clear that other products or methods would yield more useful results (e.g., summaries of statistical analysis of large ensemble data sets may provide the needed information with higher confidence than downscaling). Another useful approach could be to develop a typology of decision-making situations and information, and to accrue a repertoire of data and methods that can be adapted for similar applications. It is not that national-scale assessments are unable to meet the need for tailored knowledge; they simply cannot cover all needs at once, and must develop strategies for managing the tension between being nationally comprehensive and providing information that is detailed enough to meet adaptation and mitigation decision-making needs at sub-national-to-local scales.

4.3 Tension between scope, complexity, and manageability

There are benefits to creating NCA syntheses through an inclusive process, involving a large number of people. The NCA3 process involved a 60-members NCADAC, more than 240 authors, and around 1,000 additional contributors to underlying foundational and technical input documents. To some, this seemed excessive – but the benefits of broad engagement were visible throughout the process.

Participants noted the richness of experience brought by the wide array of authors, the capacity to deal with diverse process and topical issues, and especially the virtues of a distributed network that could be mobilized for different events. Having such a large and diverse team meant it was relatively easy to engage people in local areas and sectors during the report release and subsequent outreach. However, federal agencies were legitimately concerned about the cost and complexity of such extensive participation throughout the process, and whether this level of effort could be sustained.[2] Even with a mostly volunteer "army" working on the report, the costs associated with managing the process must be acknowledged and objectively evaluated relative to the benefits.

As described in more detail in other contributions to this Special Issue (Jacobs and Buizer 2015 this issue; Cloyd et al., Submitted for publication in this special issue), the three completed national assessments took different approaches in this regard. The first NCA (NCA1) shared with NCA3 an emphasis on engagement in which stakeholders provided input about their information needs and learned from scientists about climate change. This direct engagement in NCA1 raised expectations among stakeholders of further engagement, and then lead to disappointment because no resources were available to sustain these nascent relationships.

[2] For the NCA3, NOAA paid for federal advisory committee activities, some author travel, and the NCA technical support unit from a budget line established for this purpose by the Office of Management and Budget, but many other USGCRP agencies contributed staff and resources.

The second NCA (NCA2) was efficient and produced a credible document, but – based on the very limited press coverage relating to NCA2 and feedback received from a wide range of participants in NCA3-related events – failed to reach the American public in a meaningful and sustained way. The tension here is between the benefits of big vs. small synthesis efforts, and the benefits of an easily managed process vs. the benefits of a broad engagement strategy. In the NCA3, assessment leaders took the more engaged, if more complex, approach in light of growing stakeholder expectations and needs for salient information to inform mitigation and adaptation efforts.

As described in Buizer et al. (2013), multiple advantages can be derived from a highly strategic and well-managed sustained assessment process. From the perspective of conserving resources at the federal level, a key benefit of a more distributed process (in which self-motivated users work with data, products, and tools) is that it shifts some of the assessment burden on to intermediaries (e.g., in the NGO or private sectors) and entities conducting their own assessments. This helps to avoid burn-out of the relevant expert community and increases the human capital for doing assessments (i.e., the core ingredient of "people"), and improves the quality and utility of assessment products. A sustained assessment thus must: (1) retain (and enhance) value, utility, transparency, and credibility; (2) serve information needs of decision-makers at multiple scales, i.e., develop salient climate information; and (3) build an ongoing, manageable and legitimate process. Clearly, there will be a need to engage partners outside of government to achieve these goals, given the significant limitation on government resources.

A critical concern about reducing the overall size of the NCA effort is the need to ensure that a federal advisory committee (required when there is ongoing engagement of non-governmental participants in government processes) is constituted in a way that optimizes the utility of assessment outcomes, while adequately representing regional and sectoral interests. This is significantly more challenging than assessing physical climate changes. This tension between inclusiveness and efficiency needs to be addressed to ensure that the process supports the goals of the sponsoring program (in this case, the USGCRP) as well as the broader public interest.

Generating enthusiasm for smaller, targeted assessment products (even though those might go into greater depth, provide input to the national assessments, or fill specific knowledge gaps) may be harder than getting people engaged in a big, highly visible national effort every 4 years. In aggregate, multiple shorter, targeted products and synthesis reports could even be more expensive than the NCA3 approach if not carefully managed. On the other hand, there are substantial benefits and satisfaction from meeting user needs. The attraction of working on a national assessment that would become *the* foundational scientific document for executive branch activities was a significant incentive for the authors of the NCA3. How can a more distributed, ongoing and evenly-paced process generate the same level of commitment? This will be a significant challenge, given that resources will be needed from both federal and non-federal participants. Assessments can certainly be done more cheaply, but getting useful outcomes per dollar spent is really the goal.

If a sustained process can be implemented quickly, a "bottom up" meets "top down" sustained assessment approach that involves external partners with regional and sectoral teams and incorporates a federal structure and management team seems like an obvious next step based on the lessons learned and capacity built during the NCA3. Building from this capacity to a long-term approach that facilitates contributions from external parties within a federal framework will help to match resources with expectations.

4.4 Tension between deliberate evaluation and ongoing learning approaches

The NCA3 leadership decided from the start to build ongoing monitoring and evaluation into its operations. This was not only evident in its strategic plan and the transparency and effectiveness goals, but also by making the framing and approach to "evaluation" an explicit assignment of a NCADAC Working Group.[3] At first, the whole idea of evaluation was resisted, in part because evaluation was seen as a task separate from assessing the science and delivering a report, and in part because it was interpreted as passing judgment on performance, which could backfire and undermine agency and funding support for the entire enterprise, or open the door to public criticism.

This led the evaluation working group to suggest framing the NCA as a "learning organization," and make evaluation an ongoing and useful tool to enable participants to make course corrections as needed. In addition, mechanisms were established to track activities, outputs, impacts, and to collect feedback during regional town halls and other workshops. These formal tracking mechanisms and other informal mechanisms including "reflection time" during NCADAC meetings (and its Executive Secretariat) and NCA staff meetings, fostered flexibility and learning internally. For example, a staff-led analysis of media coverage after the release of the public review draft helped support the rollout strategy and provided encouraging feedback to the authors and NCAnet, further building social capital among volunteer partners. After the completion of the NCA3, the sponsoring agency (NOAA) also administered a short survey to NCADAC members on the experience of participating in the assessment for its own internal purposes and then USGCRP supported a process workshop focused on lessons learned in the NCA3. At the workshop evaluation experts and participants came to general agreement on what an external evaluation could accomplish and how it could be conducted (USGCRP 2014), but as yet no funding has emerged to support a formal external evaluation. It is unclear whether federal agencies will see the potential value of such evaluations as part of the sustained assessment process to be greater than the potential risks, or whether a foundation or research funding agency will have enough interest to support such an effort.

Thus, the tension over when or how to build monitoring and evaluation into assessment processes has remained focused on internal, ongoing, learning-oriented mechanisms. Our experience shows that this is an absolute necessity. But without external, independent eyes, future assessment designers lack an essential feedback mechanism. We view this as a critical opportunity missed that could result in less efficient future investments in assessment activities.

5 Conclusions

The design and conduct of assessments always involves tensions between what is scientifically defensible and practically feasible. Assessments can – if done well – have important policy, economic, and regulatory implications while providing important milestones for the scientific community. These contexts and tensions need to be explicitly recognized and managed in order to ensure that the internal management and conduct of assessments are feasible and

[3] Because the evaluation task was later determined to be too large to be included in the work of the Engagement and Communication working group, the evaluation task was transferred to the working group focused on developing recommendations for the sustained assessment process.

productive, while the process and its outputs and outcomes are perceived as credible, salient, and legitimate. The widespread frustration among climate scientists and citizens associated with the inability of the US federal government and previous administrations to effectively address the causes and effects of climate change is probably one key explanation for the strong interest among NCA3 participants and contributors in volunteering to assist in the process. The groundswell of support for this effort crossed sectors and regions and included people from many walks of life, and formed the foundation for further engagement in a sustained assessment process.

The enormous investments of time, money, and human capital made in the NCA3 resulted in considerable good will, great appreciation by both insiders and outsiders of the process and products, and in significant growth in multi-disciplinary, interdisciplinary, and transdisciplinary networks of experts and assessment users. These, to us, provide a foundation and opportunity for the sustained assessment that is too good to lose. Ideally, the energy in the underlying tensions discussed here and prevalent in every assessment, if properly recognized and managed, can be harvested in strategic ways to increase learning and greater engagement of experts and users at all levels. The intent of this concluding paper, and indeed of this Special Issue, has been to articulate some of the tensions and lessons learned so that future assessment processes can launch from this fundamental understanding rather than having to re-learn these same lessons.

References

Buizer JL, Fleming P, Hays SL, Dow K, Field CB, Gustafson D, Luers A, Moss RH (2013) Report on preparing the nation for change: building a sustained national climate assessment process, national climate assessment and development advisory committee, Washington, DC

Buizer JL, Dow K, Black ME, Jacobs KL, Waple A, et al. (2015) Building a sustained climate assessment process. Climatic Change. doi:10.1007/s10584-015-1501-4

Cash DW, Clark WC, Alcock F, Dickson NM, Eckley N, Guston DH, Jäger J, Mitchell RB (2003) Knowledge systems for sustainable development. Proc Natl Acad Sci U S A 100:8086–8091

Clark WC, Mitchell RB, Cash DW (2006) Evaluating the influence of global environmental assessments. In: Mitchell RB, Clark WC, Cash DW, Dickson N (eds) Global environmental assessments: information and influence. The MIT Press, Cambridge, MA, pp 1–28

Cloyd ET, Moser SC, Maibach EW, Maldonado JK, Chen T (submitted for this issue) Engagement in the third US national climate assessment: commitment, capacity, and communication for impact. Climatic Change

Executive Office of the President (2013) President's climate action plan. June 2013

Galford GL, Betts AK, Carlson S, Ford S, Hoogenboom A, et al. (submitted for this issue) Opportunities and strategies for statelevel climate assessments. Climatic Change

Global Change Research Program (USGCRP) (2014) Frameworks for evaluating the national climate assessment: workshop report. USGCRP, Washington, DC

Grimm N, Groffman PM, Staudinger MD, Tallis H (submitted for this issue) Climate change impacts on ecosystems and ecosystem services in the United States: Process and prospects for sustained assessment. Climatic Change

Howden M, Jacobs KL (submitted for this issue) Innovations in assessment and adaptation: building on the US national climate assessment. Climatic Change

Jacobs KL, Buizer JL (2015) Building community, credibility and knowledge: the third US National Climate Assessment. Climatic Change. doi:10.1007/s10584-015-1445-8

Kenney MA, Janetos AC, Lough G (submitted for this issue) Building an integrated national climate indicator system, Climatic Change

Liverman D (2015) U.S. national climate assessment gaps and research needs: overview, the economy and the international context. Climatic Change. doi:10.1007/s10584-015-1464-5

Maldonado JK, Bull Bennett TM, Chief K, Cochran P, Cozzetto K, et al. (submitted for this issue) Engagement with indigenous peoples and honoring traditional ecological knowledge, Climatic Change

Miles EL, Snover AK, Whitely Binder LC, Sarachik ES, Mote PW, Mantua N (2006) An approach to designing a national climate service. Proc Natl Acad Sci U S A 103:19616–19623

Morgan MG, Cantor R, Clark WC, Fisher A, Jacoby HD, Janetos AC, Kinzig AP, Melillo J, Street RB, Wilbanks TJ (2005) Learning from the U.S. national assessment of climate change impacts. Environ Sci Technol 39(23):9023–9032

Moser SC, Finzi Hart J (2015) The long arm of climate change: exploring climate change impacts via teleconnections. Clim Chang 129(1–2):13–26

Moser SC, Davidson MA (2015) The third national climate assessment's coastal chapter: the making of an integrated assessment. Climatic Change. doi:10.1007/s10584-015-1512-1

Moss RH (submitted for this issue) Assessing decision support systems and levels of confidence to narrow the climate Information "usability gap". Climatic Change

National Research Council (NRC) (2007) Analysis of global change assessments: lessons learned. National Academies Press, Washington, DC

Waple AM, Champion SM, Kunkel KE, Tilmes C (submitted for this issue) Innovations in information management and access for assessments. Climatic Change

MIX
Papier aus verantwortungsvollen Quellen
Paper from responsible sources
FSC® C105338

If you have any concerns about our products,
you can contact us on
ProductSafety@springernature.com

In case Publisher is established outside the EU,
the EU authorized representative is:
Springer Nature Customer Service Center GmbH
Europaplatz 3, 69115 Heidelberg, Germany

Printed by Libri Plureos GmbH
in Hamburg, Germany